ETHEREUM FOR BEGINNERS

The Ultimate Guide to Mining, Trading, and Investing in Cryptocurrency

(The Ultimate Guide to Ethereum and Ethereum Mining)

Pamela Kinca

Published by Tomas Edwards

Pamela Kincaid

All Rights Reserved

Ethereum for Beginners: The Ultimate Guide to Mining, Trading, and Investing in Cryptocurrency (The Ultimate Guide to Ethereum and Ethereum Mining)

ISBN 978-1-990373-66-4

All rights reserved. No part of this guide may be reproduced in any form without permission in writing from the publisher except in the case of brief quotations embodied in critical articles or reviews.

Legal & Disclaimer

The information contained in this book is not designed to replace or take the place of any form of medicine or professional medical advice. The information in this book has been provided for educational and entertainment purposes only.

The information contained in this book has been compiled from sources deemed reliable, and it is accurate to the best of the Author's knowledge; however, the Author cannot guarantee its accuracy and validity and cannot be held liable for any errors or omissions. Changes are periodically made to this book. You must consult your doctor or get professional medical advice before using any of the suggested remedies, techniques, or information in this book.

Upon using the information contained in this book, you agree to hold harmless the Author from and against any damages, costs, and expenses, including any legal fees potentially resulting from the application of any of the information provided by this guide. This disclaimer applies to any damages or injury caused by the use and application, whether directly or indirectly, of any advice or information presented, whether for breach of contract, tort, negligence, personal injury, criminal intent, or under any other cause of action.

You agree to accept all risks of using the information presented inside this book. You need to consult a professional medical practitioner in order to ensure you are both able and healthy enough to participate in this program.

Table of Contents

INTRODUCTION 1

CHAPTER 1: ETHEREUM 3

CHAPTER 2: SMART CONTRACTS 8

CHAPTER 3: CRYPTOCURRENCY 101 15

CHAPTER 4: ETHEREUM AND SMART CONTRACTS 37

CHAPTER 5: HOW TO MINE ETHEREUM 55

CHAPTER 6: FREQUENTLY USED TERMINOLOGIES AND THEIR DEFINITIONS 64

CHAPTER 7: SMART CONTRACTS AND ETHEREUM INTEGRATION 91

CHAPTER 8: HOW TO MINE CRYPTOCURRENCY 98

CHAPTER 9: THE ETHEREUM ENTERPRISE ALLIANCE 112

CHAPTER 10: USES OF ETHEREUM 116

CHAPTER 11: HOW TO BUY, SELL AND STORE ETHEREUM 122

CHAPTER 12: HOW MARKETS ARE CHANGED BY CRYPTOCURRENCIES 128

CHAPTER 13: ROADBLOCKS 133

CHAPTER 14: TIPS ON INVESTING IN ETHEREUM 135

CHAPTER 15: HOW TO MINE ETHEREUM 156

CHAPTER 16: USES OF ETHEREUM 162

CHAPTER 17: WAYS TO MAKE MONEY WITH ETHEREUM 166

CHAPTER 18: WHAT IS ETHEREUM PROOF OF STAKE..... 179

CHAPTER 19: PRIMARY PURPOSE OF ETHEREUM 184

CHAPTER 20: CRYPTOCURRENCY SECURITY 188

CONCLUSION 195

Introduction

"Like most of you reading this book, I'm just a regular guy with a regular day job. I began my investing career, more of a hobby really when I got my first paycheck at age 15 and opened my first bank account. I've expanded my horizons as I've gone through life, especially with a wife, a mortgage, and kids.

I believe in "smart investing" – this means no get-rich-quick schemes but also not passively letting your money sit idle in a checking account or your grandmother's mattress. A smart investor studies before executing any potential trade and understands the risks and rewards that come with investing. I developed an interest in cryptocurrencies after learning of Bitcoin several years ago. I was fortunate enough to "get in" early in the game and sold many of my Bitcoins at a handsome profit, though I still hold a fair number as well.

My knowledge of Bitcoin lead me down the path of other cryptocurrencies before I came across the Ethereum platform. As I will explain in the book, Ethereum sets itself apart from other similar networks through its use of smart contracts. This is why I believe that Ether, the cryptocurrency that Ethereum requires to operate, is a smart investment to add to a well-diversified portfolio.

This book should NOT be construed as just plain investing advice. I am in no way advocating that you should throw your life savings in cryptocurrencies or raid the retirement fund for this purpose. You may even finish this book and decide that Ether is not for you, and that's completely fine. This book is meant to be a guide and a starting point to enter the exciting world of cryptocurrency investing. It begins with a background on what cryptocurrencies are, then is followed with a description of Ethereum. I hope you find this book as useful as I found it fun to write!"

Chapter 1: Ethereum

If you haven't heard of Ethereum, then do not worry because this book is here to help. Ethereum is a form of crypto-currency. Crypto-currencies are also known as digital money. The 'newest kid on the block' is also the number two kid on the block now: Ethereum. Ethereum has its own crypto-currency tokens, named Ether. In the crypto-currency market, it is ranked second after Bitcoin tokens with regards to the market value and the volume of tokens transferred daily. To understand how broad Ethereum is, let us start from a more detailed definition.

Ethereum is an open source network that is built on blockchain technology and employs the use of smart contracts to run transactions in the system. Blockchain technology is, in a way, much like the internet with regards to robustness. The blockchain is a public ledger which allows information to be stored continuously. Smart contracts, on the other hand, are automatic, self-executing contracts on the internet. The terms of these contracts are written in the lines of the code itself.

Ethereum is, therefore, a 'world computer'. Every computer in the network works both as a node or connector and, also, as a server, but we will delve into those details later on in the book.

Though Ethereum is often compared to Bitcoin, it goes further than just offering exchange and storage of its crypto-currency, Ether. The platform was created to give developers a non-censored, decentralized system. Here, they could build applications to serve the needs of many different industries and not just the world of crypto-currency.

Blockchain technology makes Ethereum a very broad programming network. A developer can build social apps, financial apps, make games and virtually everything we use in today's world. Blockchain technology allows seamless decentralization of information on the network. There is no central authority controlling all the information running the applications on the system.

History

The story of Ethereum cannot be told without going back to Bitcoin. This does not mean that they are the same. So, just how did this Ethereum computer come to be? Well, many years ago, before the creation of Bitcoin, many developers had attempted the creation of different crypto-currencies or online 'digital money'. Their efforts were successful, but they all had a common flaw. There was the issue of double-spending. This issue of double-spending was when the tokens on the network, much like Bitcoin or Ether, would be spent twice.

In the year 2008, a person—or group of individuals—under the pseudonym Satoshi Nakamoto, published a white paper called Bitcoin: A Peer-to-Peer Electronic Cash System.[1] This white paper became a game changer in the cryptocurrencies market.

The network was, soon after, launched on January 3rd the following year. Many Bitcoin 2.0 projects were then started to build off the original code. The developers, however, had to rely on the Bitcoin system or protocol and were finding it especially hard. A Russian programmer and designer, Vitalik Buterin, was one such developer. Vitalik found that incredibly talented and skilled developers were struggling to build off the code while remaining within the security of the protocol. He, therefore, took a step back and decided to come up with a platform that made it easier for developers to create decentralized apps while remaining within the full security of the protocol.

Therefore, in late 2013, he published the proposal for such a platform, which he named Ethereum. The development of Ethereum was funded by a crowd sale that occurred in mid-2014. On 30th of July 2015, the platform was successfully launched. $11.9 million worth of Ether was pre-mined before the launch.

Due to a hack that occurred in 2016 on the DAO project, Ethereum was forked into two separate blockchains. We will take a closer look at the DAO project in a future chapter of the book.

Chapter 2: Smart Contracts

While smart contracts have the potential to change the world in a wide variety of ways, as of 2017 they are primarily seeing use among the decentralized applications found in the Ethereum blockchain. These applications are already being used for a wide variety of purposes including things like making contract negotiations much more manageable and simplifying the way in which insurance policies are paid out on vehicles that are already connected to the internet of things. This, in turn, dramatically decreases the time that is required to process a claim and requires no claims adjuster to get involved besides.

In this scenario, an insurance company would write out a smart contract as part of the customer' s policy along with any other specific conditions that may be required. Then, if any of these conditions are picked up by the vehicle, the information is noted by the smart contract and the claim is paid.

Smart contracts are also already making significant strides when it comes to the way in which copyrighted content is managed. The company Blockphase is already marketing a tool that can help content creators who work in the augmented reality, virtual reality or 360-degree video space ensure that their content remains secure while online. Content they create is then uploaded into Blockphase's blockchain and it is then automatically searched for continuously. If a match is found, Blockphase then initiates a smart contract to the user of the illegal content to request payment. The end result is that artists are able to more easily focus on being artistic and less on ensuring that they are being paid for their content.

When it comes to digital media, the company ContentKid is working on changing the way that content subscriptions are dealt with. The way it works is that it creates a large number of accounts across services like Hulu and Netflix. Using the Ethereum blockchain and smart contracts these accounts are then rented out to individuals in chunks based on the type of content that is going to be streamed. Thus, users can pay for a half an hour of time if they are looking to catch the latest episode of their favorite sitcom or pay for 13 hours of time to binge a whole season of a scripted drama. Payment for the service happens automatically over the Ethereum blockchain as smart contracts automatically charge users as they watch.

Great strides are also being made when it comes to cloud storage through companies like Storj Labs. This company is using the Ethereum blockchain and smart contract technology to link together people who are interested in additional cloud storage options with those who have the hard drive space to spare. Their app takes advantage of the inherent security available thanks to blockchain.

Shaping the future

While these are just a handful of the decentralized application options that are currently making good use of blockchain and smart contract technology today, where smart contracts are going to really shine is when it comes to shaping the future.

When it comes to capital markets, interest in blockchain and smart contract has already been expanding for multiple years, with investment interest in this sector seeing 100 percent growth in 2015, and then again in 2016. This is largely the case as previous advances in this space have been more focused on front-office activities while leaving the middle and back offices to wait their turns, content to get along with outdated and frequently inefficient practices.

The Linux Foundation is currently working on a way to change all that and is striving to bring together a mix of capital market firms along with existing blockchain companies for the purpose of developing standards and creating smart contract and blockchain technology that is specifically useful to the capital markets. Instead of changing this market completely, it is more likely that smart contract and blockchain technology are going to work together in such a way that it will help to suss out much of the remaining inefficiency from the system.

When it comes to real estate, the transactions that are required in order to buy or sell a home are incredibly tedious and painstakingly monotonous, largely because the industry hasn't updated these practices since the invention of the internet. Luckily, smart contract and blockchain technology are poised to do wonders for this industry and the entire listing process can be handled via smart contracts, and they will also be able to cut through much of the red tape with the entire transaction, largely cutting real estate agents out of the loop entirely.

By breaking down the traditionally centralized structure, the system will naturally provide those in the real estate industry with a greater degree of access to a wider variety of fee structures that will give them more control over their business than ever before. What's more, once the system is based out of the Ethereum blockchain, listing will be easily visible to anyone with access to the chain which means buyers will always have access to the latest data and sellers will be able to reach a greater portion of the market than ever before.

When it comes to the public works sector, it is an extremely complex system in that it is centralized when it comes to its responsibility in carry out the delivery of public services while also remaining fragmented in the way that each service is actually carried out and how various departments share data. The effects of this duality typically cut deep as departmental budgets are slashed and questions arise over decreasing services or changing the way in which they are delivered.

As smart contract and blockchain technology becomes more mainstream, however, it will be useful when it comes to weeding out the inherent inefficacy in the system. When given the chance, smart contracts can be used to make the payment and issuing of government license much more streamlined and it can be used to create an official registry that monitors all of the intellectual property the agency might own. It can also serve to make coordinating the purchasing process much more manageable, ensuring that limited government budgets can be stretched as far as bureaucratically possible. In each of these instances it is clear that smart contract and blockchain technology will decrease response times, while at the same time decreasing the possibility of error and fraud while also increasing productivity at the same time. In a matter of speaking, anywhere that you find government inefficiency, smart contracts can be brought in to sort it out in short order.

Chapter 3: Cryptocurrency 101

If you feel like you're swimming through mud a little at this point, by the end of this chapter, you will be swimming through crystal clear water!

This chapter is going to give you a lot of information, but it is vital to allow you to really understand the subject at hand. It might be useful to grab a pen and paper and make notes as you go long, as we are going to cover a lot of ground, with many useful hints and tips. Don't worry if this is a little overwhelming at this point, and if you still have any questions by the end of this book, they will surely be covered in our comprehensive final chapter, with all the FAQs to do with cryptocurrency.

Okay, so let's begin.

We are going to assume at this point that you are a total beginner, that you know virtually zero in-depth about cryptocurrency, and that you are simply interested and intrigued by the whole thing. Starting with a clean canvas is the best situation to be in, as you can learn and develop as you go. If you do have any prior knowledge, make sure that you continue to read all the chapters, without skipping, as you may learn many new titbits to see you through your journey to cryptocurrency master.

A Detailed Chat on What Cryptocurrency Essentially is

In our introduction section we briefly mentioned that cryptocurrency is money, a currency. The difference between regular cash and cryptocurrency is that you will never hold cryptocurrency in your hand, unlike a dollar or a pound, for example. Cryptocurrency is digital, it is all in the ether, and cannot be held or seen.

This might lead you to believe that because you can't see it or hold it, it isn't worth anything – wrong!

You can trade cryptocurrency for regular cash, you can invest in it, and it can lead to a rather lucrative future, if done the right way. Put simply, any type of cash is an entry on a database. Whilst you can hold cash in your hand, the actual workings behind it adhere to a paper trail. For instance, your bank account statement would tell you exactly where you obtained money from, where you spent it, how much you spent, who you sent cash to, etc. No transaction is able to be made without an action, e.g. you drawing cash out of an ATM, or you sending money online. Cryptocurrency is no different.

Cryptocurrency is a series of database entries with monetary value, and instead of a bank statement, there is something called a public ledger. We will go into much more detail a little later on, in terms of how it all works, but this public ledger basically tells everyone where the money is, where it went, who sent it, and ensures that it is never spent twice.

The basic characteristics of cryptocurrency are:

• Decentralised – We're going to cover this in more detail shortly, but cryptocurrency is not under bank ruling, and is therefore decentralised

- Cannot be used twice
- Private and confidential
- All transactions are done electronically

That's really all there is to know about the basic characteristics of a cryptocurrency. There are many different types of cryptocurrency, which could loosely be referred to as 'brand names', such as Bitcoin, for example, and we're going to go into much more detail on the main ones in a later chapter.

You might be wondering 'why bother?' Well, in that case, the advantages of cryptocurrency are:

- The fact that cryptocurrency is decentralised means that it doesn't fall foul of economic downturn, making it more stable overall for investment purposes
- Lower fees for transferring cash, if any fees at all

- Fast transfers worldwide. You're not going to have to wait x amount of working days for an international transfer to complete, it will be instant or just take a few minutes

- Very little chance of fraud

- No chance of identity theft

- You have total control over your cash

The Differences Between Regular Cash and Cryptocurrency

There are of course differences between regular cash and digital currency, and the advantages above really spell out the biggest ones. Decentralisation is probably the one that makes the biggest difference to users.

The economy, both nationally and globally is affected by the smallest thing, and that can mean that your investment is prone to both upturns and downturns. Whilst upturns are great news, downturns are pretty bad; there is probably more chance of a downturn these days, thanks to all manner of funky things going on in the world. In that case, you are never secure in your investment, which can be very off-putting to the first time investor in particular. With cryptocurrency, your investment is not going to be affected by someone pulling out of the EU, or by a drop in oil exports – the only affecting factor is cryptocurrency and its popularity itself. This is on the rise in a big way, so it's really all good news.

The decentralisation issue goes further that than too. When you send cash overseas, perhaps to family or friends, then you usually have to wait for around a week, and there are quite hefty fees to pay. Even if you choose to send cash via an app, such as PayPal, the fact it is centralised means that you are going to have to pay a percentage fee of that transfer to PayPal itself. This is not the case with cryptocurrency. Transfers take literally minutes maximum, and there are no fees involved. A transaction with cryptocurrency is exactly the same whether you are transferring money to China or someone next door to your home.

Another difference, and one of the biggest plus points, is that the chances of identity theft by using cryptocurrency is virtually impossible. When you use your credit or debit card to pay for anything online, you are at risk of someone stealing your details. Whether you use a secure site or not when paying online, there are ways for hackers to get in there and cause you all manner of costly problems. With cryptocurrency, this isn't a likelihood or even a possibility.

When you carry out a transaction with cryptocurrency, you use your own personal 'key'. Nobody else knows this key, it cannot be unlocked and it cannot be traced. Consider a cryptocurrency transaction to be like Superman – literal steel, nothing is getting through it.

Whilst if you pay for something using a credit card and there is an issue, you do have the protection the card service offers, this can sometimes involving jumping through hoops to activate, and can take a lot of time for the cash to arrive back in your account, or on your original card.

The History of Cryptocurrency, and How it Has Developed

You don't have to know anything about cryptocurrency to have heard of Bitcoin. This is the first and most famous of all the digital currencies, and is really what started the whole trend off.

Bitcoin was developed by Satoshi Nakamoto in 2008, and steadily grew to be a large monetary force. After seeing success, other people and companies began to catch onto the trend, with other digital currencies being forged. Litecoin, Altcoin, and Ethereum were all born off the back of Nakamoto's brainwave, and there are countless more besides. You can even set up your own cryptocurrency if you want to!

At first people didn't really understand Bitcoin, and it really flew under the radar for a little while, ebbing and flowing, but keeping enough steady momentum going to add up to a snowball effect. Nowadays, Bitcoin is one of the most commonly used payment methods for online goods and services, with more and more vendors signing up to take the currency. That means you could pay for your cosmetic surgery in New York online, by choosing to pay by either credit card, debit card, PayPal, bank transfer, for Bitcoin.

Serious progress in a short space of time, we think you'll agree. Much of the growth in popularity is down to education; more of us now know what this is all about, and as a result, we can understand the benefits much better than before.

Where The World of Cryptocurrency Can Take You

There are really four main usage routes when it comes to cryptocurrency:

- Trading

- Investing

- Using

- Creating your own

Each of these routes is part of our 7 secrets, the whole point of this book, so we are going to dedicate a chapter to them later on. For now however, you simply need to know that trading and investing in cryptocurrency can be a lucrative side-line; using cryptocurrency is a quick, cheap, and useful way to pay for goods and services, and send transactions to various other users, and creating your own cryptocurrency could be a very useful money making scheme, once you have all the finer details of cryptocurrency down pat, and once you have experience.

Can You Really Make Your Own Cryptocurrency?

Yes! And again, we will talk about this in more detail later on in our book. Basically, you need experience when it comes to creating your own version of cryptocurrency, and you need patience.

There is a very real chance your cryptocurrency won't really create much revenue, but there is also a chance that if you can get enough miners on the job (more on that later), then you could steadily push up to Bitcoin levels!

Yes, it's going to take time, but Nakamoto started at zero once upon a time!

Let's Get Specific – How Does Cryptocurrency Actually Work?

Okay, we've talked at length about what this whole world is about and we know where it all came from, but how does it actually work?

In order to fully understand and utilise something, you need to know the process of how it actually works. You don't need to go into algorithm specifics, because that really will send you to sleep, but you do need to know the basics of how a transaction works.

Let's break it down into easily digestible bite-sized chunks.

Firstly we will assume that you already have an account (known as a 'wallet') with that particular cryptocurrency, i.e. you will have a Bitcoin wallet, and the person to whom you are sending cash, or the company you are paying, also have a wallet. This makes sense, because you literally transfer the funds from wallet to wallet. This doesn't have to be Bitcoin. Remember, it can be any cryptocurrency, but you need to have like for like 'wallets' in order for the process to work.

Signing up for a 'wallet' is free, and it takes a very short space of time to complete the process. You do need to make sure you check out which cryptocurrency is best for you, before choosing one however. This is something we will cover in a later chapter.

So, the process goes a little like this.

Person requests for a transaction to take place. E.g. Colin decides he wants to send funds to Dorothy. Both Colin and Dorothy have wallets for the same cryptocurrency. Colin therefore goes online and requests for the transaction to take place, entering Dorothy's details, and using his unique 'key' to complete the request.

The request for the cash to be sent from Colin to Dorothy is received by the cryptocurrency server. This is all done super-fast, by the way. The server is a network of computers which are collectively called 'nodes'.

The nodes first ned to validate the request for the transaction, before they will allow it to go through. This is why cryptocurrency is so secure. Those confusing algorithms that we won't bore you with are used at this stage, and it basically validates Colin's key and request against each other.

When the nodes have decided that all is well with Colin's request, and that he is who he says he is, a new entry or data block is created on the ledger. The ledger is the same as your bank statement in regular terms, and records every single transaction, ensuring that money is not spent more than once.

The block chain is updated and it is unable to be changed from this point onwards; it is set in concrete and nobody has the authority or power to make a change. Again, this is about security and this is why it is almost impossible to forge a cryptocurrency transaction. Basically, there's no getting past those pesky nodes.

The funds are sent from Colin's wallet to Dorothy's wallet, signed off electronically with Colin's key. The ledger will now state that user X sent funds of such and such amount to user Y. Dorothy is happy, she now has cash, and Colin can rest easily that the transaction is a successful one.

This is all done within a blink of an eye and costs zero cash to complete.

Basically, the ledger is the holy grail of the cryptocurrency we are talking about. This is the record of every single transaction that has gone on with that particular type of cryptocurrency, and whilst it is decentralised, the ledger is what holds it all together, the glue if you will. Without the ledger, all hell would break loose, because then it would be possible to spend a unit of cryptocurrency more than once, and that is the underpinning law of any currency, whether regular or digital.

The Risks of Cryptocurrency

Everything in life has risks to balance up against the good points, and in the interests of being correct, we need to talk about the risks that can be associated with cryptocurrency use of any kind. The risks are low, but they are there too. Basically, the risks are more to do with investing, because investment by its very nature is a risk to some degree.

Everything is Digital

The biggest risk in terms of using cryptocurrency is the fact that it is in the digital realm. When you can't hold something in your hands, you feel that you can't control it. Of course, the same could be said for your savings in a bank account, because they are just numbers on a screen too, but the chance of a hacker or malware attack taking your coins is a real risk to think about.

This is why you should research your particular cryptocurrency very carefully indeed, and invest in some high quality Internet security for your primary device. Remember, if you are going to be checking your wallet or doing any form of transaction on the go, you need to protect your smartphone or tablet too. This will go a long way to reducing the risks, but we can never completely rule out the chances of a large cyberattack.

You May Lose Your Password

What if you lose your password and you become locked out of your wallet? You're screwed basically. For that reason, just don't forget it, but don't write it down anywhere either.

Using a Bad Exchange

We will talk in our next chapter about the use of an exchange when buying and trading cryptocurrency,you will see that there are countless out there. Because there are some unscrupulous folk on this planet, it is important to know that the exchange you're using is reputable. A dodgy exchange could literally screw you right over.

Making a Bad Investment

If you are investing for the first time in cryptocurrency, it is best to start small. There are some great investment opportunities out there, but if you are a little 'wet behind the ears', you could make a loss before you even begin. Check out every opportunity before you invest, and work your way up to the big ones. Never run before you can walk in any investment situation, whether cryptocurrency or otherwise. Just because you can't physically see cryptocurrency, doesn't mean it isn't cold, hard cash with value.

Panic Selling and Taking a Big Loss

We mentioned that cryptocurrency isn't as affected by ups and downs in the market, because it is decentralised, and that is true to a very large degree; not entirely however. Every single investment on the planet is going to go through great times, and not so great times. Whilst the ebb and flow might not be as dramatic, there is still the chance that you could see a downturn and consider it best to sell your coins quickly, acting in panic. Knowing when to wait it out and when to act is something that is learnt with experience.

The Unique Language of Cryptocurrency

The world of cryptocurrency is unique and a little confusing at first, so it makes perfect sense that there is likely to be a whole other language connected to it. We could go on to list every single term that is related to cryptocurrency buying, trading, investing, and how it all works but we would only confuse matters further. For that reason, let's talk about the terminology connected with cryptocurrency that you are likely to come across quite early on in your journey from beginner to master.

In no particular order, these are the terms that will help you decipher the original cryptocurrency language.

- Address -This is the same as a website address, and it is where you keep your cryptocurrency. Rather than a line, it is a mixture of encrypted letters and numbers.

- Altcoin – This is a collective term for any type of cryptocurrency that isn't Bitcoin.

- Block chain – With every single transaction, another chain is created, and this growing chain is called the block chain. This is data on the ledger which records that particular cryptocurrency.

- Block – An entry on the block chain. Once a block is recorded it cannot be removed.

- Mining – This is something we will cover in more detail later on, but mining is the process of figuring out blocks in the block chain, using confusing algorithms. When this is done, the person is given a mining reward, which is usually coins.

- Node – A computer in a group which validates and confirms transactions.

- P2P – This is a system that helps to create the block chain. The decentralised nature of cryptocurrency is massively aided by the P2P system, which stands for peer to peer.

- Signature – This is not a literal signature, but a mathematical method, e.g. entry of a key, which proves that the wallet and included coins are in fact yours.

- Exchange – A website from which you buy and sell various forms of cryptocurrency.

- FIAT – A form of currency which is created by the Government, e.g. the US Dollar, the British Pound.

- Whale – Someone who has an awful lot of cryptocurrency in their midst.

- Bullish – A clear expectation that there will be an increase in price.

- Bearish – A clear expectation that there will be a decrease in price.

- Pump & dump – When investing in cryptocurrency, pump & dump is a common ter. This is a cycle of an Altcoin coming super popular out of nowhere, increasing in value quickly, and then suddenly dropping.

- Market cap – The total value of a particular cryptocurrency.

- ROI – Return on Investment. This is how much profit you make compared to your original investment amount.

- Cold storage -A method you can utilise to reduce your chances of being attacked by hackers, and involves moving your cryptocurrency offline.

Of course, there are countless other terms which are associated with the cryptocurrency world, but without wanting to confuse your mind completely at this point, these are the terms that will hold meaning for you during your original venture into buying, trading, and investing.

This chapter has been a mammoth one, but all the information within it links together really well, and has been set out to be easy to follow. If you had preconceived ideas about cryptocurrency being super-confusing before, hopefully by this point it will all be much clearer. You're not expected to know it all by this stage, but you should be feeling much more excited about the prospects that lie ahead in this world that is super-technical, and super-lucrative at the same time.

Chapter 4: Ethereum and Smart Contracts

I am assuming that you have at least a basic understanding of cryptocurrencies and blockchains and how they work. If not, please learn before you come back to this. You need to understand:

Public and private keys

Why miners are needed by blockchains

Decentralized Consensus

Transactions

Concepts of scripting and smart contracts

You will also need to understand more about the EVM and gas.

The main reason for Ethereum was as a platform for smart contracts and these are run on the Ethereum Virtual Machine. This gives us a better language for scripting than Bitcoin does. Given that these contracts run in EVM, to limit the number of resources used by each contract, each operation is executed by all nodes in the network. A contract code for an Ethereum transaction triggers reads and writes of data, sends messages to other contracts, and can do some expensive computations, among other things. Each operation has a cost that is measured in gas and each of the gas units that are "eaten" by a transaction is paid for in ETH. The cost is based on a price of gas to ETH which is dynamic and constantly changing. The price is automatically deducted from the account that sends the transaction. Each transaction also has a gas limit that is bound on the amount of gas the transaction consumes. This is a safeguard against errors in programming that could deplete your account of ETH.

Setting Up the Environment

Now you know the basics, it is time to begin coding. To begin your journey of developing DApps, you must have a client that connects to the network. This is your window, giving you a view of the blockchain.

There are many compatible clients but the most popular by far is called geth. However, it is not very developer-friendly so, to begin with, start using one called testrpc

To install and run testrpc, open a command prompt and type in the following:

$ sudo npm install -g ethereumjs-testrpc
$ testrpc

When you begin developing, keep testrpc open in new terminal and do your work in another one. Each time testrpc is run, 10 addresses will be generated, each with test funds simulated for your use. Do what you like with these, use them as a learning curve; it isn't real money so you can't physically lose anything.

Solidity is the most popular smart contract writing language so that is what we will use, along with Truffle development framework. Truffle is designed to help you create your smart contracts, to compile them, test and deploy them. Let's get started. Type in this at your command prompt (note that the lines beginning with # are comments, designed to tell you what we are doing – you do not need to type in these lines:

```
# First, we will install truffle
$ sudo npm install -g truffle
```

```
# now we will set up our project
$ mkdir solidity-experiments
$ cd solidity-experiments/
$ truffle init
```

Truffle will go ahead and create the files needed for a sample project and this will include a sample token contract for MetaCoin. These contracts can be compiled just by typing 'truffle compile' at the command prompt. Next, deploy the contracts to the network that has been simulated using the testrpc node. To do this, we use 'truffle migrate':

```
$ truffle compile
Compiling ConvertLib.sol...
Compiling MetaCoin.sol...
```

Compiling Migrations.sol...
Writing artifacts to ./build/contracts

$ truffle migrate
Running migration: 1_initial_migration.js
Deploying Migrations...
Migrations: 0x78102b69114dbb846200a6a55c2fce8b16f61a5d
Saving successful migration to network...
Saving artifacts...
Running migration: 2_deploy_contracts.js
Deploying ConvertLib...
ConvertLib: 0xaa708272521f972b9ceced7e4b0dae92c77a49ad
Linking ConvertLib to MetaCoin
Deploying MetaCoin...
MetaCoin: 0xdd14d0691ca607d9a38f303501c5b0cf6c843fa1
Saving successful migration to network...
Saving artifacts...

If you are using Mac OS X, you may get an error message concerning a .DS_Store file.

Truffle gets a bit confused by these so just delete the store file.

The contracts have now been deployed to the node. See how easy it is? Let's create a contract of our own.

Your First Smart Contract

We are going to write a Proof of Existence contract; the idea behind this is to create a notary that will store document hashes as proof that they exist:

$ truffle create contract ProofOfExistence1

Now you can open that contract in your text editor (use Vim, it is the best one) and paste in this code:

pragma solidity ^0.4.2;

// Proof of Existence contract, version 1
contract ProofOfExistence1 {
    // state
    bytes32 public proof;

    // calculate and store the proof for a document
    // *transactional function*

```
function notarize(string document) {
    proof = calculateProof(document);
}

// helper function to get a document's sha256
//         *read-only function*
function calculateProof(string document) constant returns (bytes32) {
    return sha256(document);
```

We are going to begin with something that is simple but not correct and move on to a better solution. This is a Solidity contract definition, akin to a class in other programming. Each contract has a state and functions and it is very important that we can distinguish between the two function types in a contract:

Read-only -these are known as constant functions, those that do not perform any changes to the state of the contract. All they do is read, perform any computations and return the values These functions are resolved node by node locally and do not have any gas cost attached to them. These are used with the constant keyword.

Transactional – these perform changes to the state and can move funds. The changes must be reflected in the Blockchain and this requires a transaction being sent to the network and has a gas cost attached to it.

The simple contract we are going to make next can only store a single proof at any given time, using the bytes32 data type. This is the same size as a sha256 hash. The function called 'notarize' lets you store a document hash in the state variable called 'proof'. This is a public variable and this is the only way a contract user can verify whether the document is notarized.

The first thing we are going to do is deploy ProofOfExistence1 but we are going to make some changes to the migrations file so that Truffle will display the contract. Replace the contents of 'migrations/2_deploy_contract.js' with this:

```
*       migrations/2_deploy_contracts.js:
*/
```

```
module.exports = function(deployer) {
deployer.deploy(ConvertLib);
deployer.autolink();
deployer.deploy(MetaCoin);
// add this line
deployer.deploy(ProofOfExistence1);
};
```

You also have the option of deleting the lines about MetaCoin and ConvertLib because we won't be needing these again. However, you must delete the test folder as well otherwise it will all fail. To run the migration, use the following command:

truffle migrate --reset

Smart Contract Interaction

We have built and deployed the contract, now it's time to have a ply. You can send messages to the contract through the function calls and you can read the public state of the contract. To do this, we are going to use Truffle console:

```
$ truffle console
// retrieve the deployed version of the contract
```

```
truffle(default)> var poe = ProofOfExistence1.deployed()

// print the address
truffle(default)> console.log(poe.address)
0x3d3bce79cccc331e9e095e8985def13651a86004

// now we register the first "document"
truffle(default)> poe.notarize('A fantastic idea')
Promise { <pending> }

// now we get the document proof
truffle(default)> poe.calculateProof('A fantastic idea').then(console.log)
Promise { <pending> }
0xa3287ff8d1abde95498962c4e1dd2f50a9f75bd8810bd591a64a387b93580ee7

// To see if the contract state was changed correctly:
truffle(default)> poe.proof().then(console.log)
0xa3287ff8d1abde95498962c4e1dd2f50a9f75bd8810bd591a64a387b93580ee7
// The hash matches up with the one we calculated before
```

All function calls will return a promise and '.then(console.log)' is used for printing the result when the promise has resolved. The first thing to do is get hold of a representation of the contract you deployed and then store it in 'poe', a variable.

Next, the transactional function called 'notarize' is called and this involves changing the state. The result will be a promise that will resolve to a specific transaction ID and not the return of the actual function. Remember, the gas must be spent to change the state and to send transactions to the network and this is why we get a transaction ID as a promise result. In our case, we are not bothered about the ID so the promise is discarded. When you write a proper app, you will need to save it so that each transaction can be checked for errors.

The next step is to call the function called 'calculateProof' which is if you remember, a read-only function. Remember, you must use the constant keyword to mark any read-only functions otherwise Truffle gets confused and will attempt to create a transaction that executes the function. That keyword tells Truffle that we are not trying to interact with a Blockchain, only read from it and by using the read-only function, we will get the sha256 size of the document called "a fantastic idea".

Now, this must be contrasted with the smart contract state so, to check if the correct change was made to the state, we must look at the public variable called 'proof'. To obtain the value of a public variable, we call a function that has the same name and this will return a promise. In this case, the output hash will be the same so it all worked exactly as it should.

Let's move on and create a better contract because the one above only registers a single document at any one time.

Iterating Contract Code

The next step is to change our smart contract so that it will support multiple proofs. Copy your original file that is called contracts/ProofOfExistence2.sol and apply the following changes. The changes are that we are going to change the variable to an array of bytes32 type and name it proofs; then we turn it private and add in a function that checks to see whether a document has been notarized. This is done by iterating the array:

```
pragma solidity ^0.4.2;

// Proof of Existence contract, version 2
contract ProofOfExistence2 {
  // state
  bytes32[] private proofs;

  // store the proof of existence in this contract state
  // *transactional function*
  function storeProof(bytes32 proof) {
    proofs.push(proof);
  }

  // calculate and then store the proof for the document
  // *transactional function*
  function notarize(string document) {
    var proof = calculateProof(document);
```

```
    storeProof(proof);
}

//the helper function that gets a document's sha256
// *read-only function*
function calculateProof(string document) constant returns (bytes32) {
    return sha256(document);
}

// check if the document is notarized
// *read-only function*
function checkDocument(string document) constant returns (bool) {
    var proof = calculateProof(document);
    return hasProof(proof);
}

// return true only if proof is stored
// *read-only function*
function hasProof(bytes32 proof) constant returns (bool) {
    for (var i = 0; i < proofs.length; i++) {
        if (proofs[i] == proof) {
            return true;
        }
```

```
}
    return false;
  }
}
```

Now we are going to interact with the functions — remember that you must update migrations/2_deploy_contracts.js so that it has the new contract in it and then run truffle migrate --reset

```
// deploy the contracts
truffle(default)>migrate --reset
```

```
// Get the new version of this contract
truffle(default)> var poe = ProofOfExistence2.deployed()
```

```
// check for a new document – you shouldn't find it.
truffle(default)> poe.checkDocument('hello').then(console.log)
Promise { <pending> }
false
```

```
// now add the document to the proof store
truffle(default)> poe.notarize('hello')

Promise { <pending> }

// now check to see if the document has been notarized
truffle(default)> poe.checkDocument('hello').then(console.log)

Promise { <pending> }
true
// success!

// other documents can be stored and recorded too
truffle(default)> poe.notarize('some other document');
truffle(default)> poe.checkDocument('some other document').then(console.log)

Promise { <pending> }
true
```

This is a much better version than the original but it isn't perfect. Whenever we want to see if a document has been notarized, it must be iterated through all the existing proofs. A better way of storing the proofs is a map and luckily, Solidity has support for maps, known as mappings. Something else we need to change is to get rid of the comments that are marking transactional or read-only functions. This is the final version:

pragma solidity ^0.4.2;

// Proof of Existence contract, version 3
contract ProofOfExistence3 {

mapping (bytes32 => bool) private proofs;

// store the proof of existence in this contract state
function storeProof(bytes32 proof) {
proofs[proof] = true;
// calculate and then store the proof for the document
function notarize(string document) {
var proof = calculateProof(document);
storeProof(proof);
// the helper function that gets a document's sha256

```
function calculateProof(string document)
constant returns (bytes32) {
return sha256(document);
// check if the document is notarized
function checkDocument(string document)
constant returns (bool) {
var proof = calculateProof(document);
return hasProof(proof);
// returns true only if proof is stored
function hasProof(bytes32 proof) constant
returns(bool) {
return proofs[proof];
```

This works the same way as the second one that we produced above. Before you try it, update your migration file and then run the 'truffle migrate –reset' command again.

Chapter 5: How to Mine Ethereum

This chapter will investigate the depths of Ethereum mining, what it entails, and how you can get started. The best place to start is by describing the general mining process. You may have come across terms like Ethereum mining and Bitcoin mining, but do you know what mining really means?

Understanding the Mining Process

In actual sense, mining is about uncovering something precious or valuable that is buried in the ground. Digital currency mining works almost in the same way as physical mining. When it comes to cryptocurrencies, mining is about a lot of computational power, time and participation.

In chapter one, we talked about networks as we were defining the blockchain. When someone enters the network and successfully performs different computations successfully, they earn a reward in the form of a cryptocurrency.

Mining is the process of looking for those rewards by participating and solving mathematical problems and computations. If you are a miner, you use your computer hardware to perform these calculations for you so that you can get paid.

You may ask yourself why the people giving out these digital coins cannot do this by themselves. Why do they need other people to do the mining? Issuers enlist the services of participants because they cannot handle this process alone. They do not have the capacity to process these tasks on their own, so they engage people from all over the world to come in and help.

This description may sound simplistic but mining is a tedious process that takes a lot of time. If you decide to be a miner, you must devote your computer to carry out these tasks. Your computer will constantly be searching through blocks trying to solve calculations and permutations.

If you are able to solve these computations, you then submit your solutions to the issuer. The issuer will sort through your answers and reward you with digital coins. Just like actually mining for gems and precious metals, you will be mining for digital currencies by utilizing a lot of computational power.

Mining is important because it is necessary to increase the number of coins in the market. This is to enable trading to take place. Trading only becomes easier when there are enough units of a cryptocurrency circulating in the market.

Mining Ethereum

Mining Ethereum is about more than just bringing more Ethers into circulation. It is also about securing the network which creates the blocks through the different transactions. In mining, the Ether unit is fundamental because it is the fuel that fires the Ethereum network. It inspires change and development because people are always looking for a way that they can create different platforms that work well with it. This approach encourages innovation and growth.

After you mine the Ether, it is up to you to decide what you want to do with it. You can either sell it or trade it for real money. There are over 18 million Ethers issued annually and this figure keeps increasing considerably every year.

When a user completes a transaction, they have to have a "proof of work" so that they can get paid. In every block, there has to be a proof of work so that the issuers can know who completed the work. Proof of work is a system that is used to ensure that miners are genuine and not hackers trying to crash the system. A miner must prove they are genuine by devoting some computational power toward solving a mathematical puzzle.

This process is not simple because there can be very many users. The algorithm (or process) helps to determine if the proof of work is legit. This validation algorithm is called Esthash. How it works is that different tasks have their levels of difficulty that the system sets. The system controls the pressure depending on the time that the miner takes to perform a function. Through manipulating the challenge, it can then identify how much time a user needs to solve a task. When mining, the difficulty level adjusts itself slowly, and there is approximately one block released every 12 seconds.

If you want to mine, you can even start in your home, but you need to have some knowledge about command prompts and script writing. Though it may look and sound like a complicated process, the secret is to break it down into smaller tasks. The process also consumes a lot of electricity, so expect high electricity bills. Nevertheless, you do make a profit when you sell your Ethers. There are online calculators that can help you determine how much profit you have made after subtracting your costs from income.

The Mining Process

The first step in Ethereum mining involves buying a computer. Get a graphics card (GPU) that has a RAM (Random Access Memory) of more than 2 GB. One word of advice is that a GPU is better than a CPU (Central Processing Unit) because it is faster for this type of mining. Other alternatives are Nivida and AMD cards. You will also need to have a lot of space on your computer for you to run the software and the different processes.

Now that we know what we need, let's look at the process itself:

Download Geth - Geth is software that will link you to the Ethereum network and allow you to communicate with issuers. It will keep you up to speed with what is happening on the network. You need to familiarize yourself with how it works before you begin.

Run the zip file - Since Geth usually comes as a zip file, unzip it into your hard drive. It is recommended that you store it in Drive C.

Search for CMD (command prompt) - Execute the application.

Locate Geth - Locate Geth on your computer.

Account creation - Type "geth account new" and press Enter so that it now becomes 'C:\>geth account new.'

Get password - Write down your password and save it. Type it in and create your new account.

Download the blockchain and link to the network - In the terminal, type 'geth —rpc' and press Enter. This action will help you get a link to the network, which may take a very long time. Be patient and wait for it to complete the harmonization before you start the mining process.

Get mining software - This will help your graphics card to work with the network. A software that you can use is Ethminer. Many people have used it successfully. Install the software that you intend t0 use.

Locate new software - Locate Ethminer or the software of your choice in your computer.

Complete the next steps - Type "cd prog" and press Tab. You will see the following 'C:\>cd prog.' Press Tab again to show 'C:/> cd "Program Files". After this, press Enter to reveal 'C:\Program Files>.'

In the software, type "cd cpp" then press Tab and Enter – You should see 'C:\Program Files\cpp-ethereum>.'

Type "ethminer – G" and press Enter to start the process - You will see a graph called the DAG (Directed Acyclic Graph). Again, make sure you have enough space on your computer to aid you in this process.

Using the CPU - You can still mine with your CPU by typing "Ethmining" and pressing the key.

Do not expect the mining to be easy. It is a long process that will take a lot of back and forth. The profits will be minimal in the short run because the initial and running costs are high. However, after some time, you will begin to make a profit from mining.

By following all the steps in this chapter, you can start your mining process on the Ethereum network. This process may make you some money, but ensure that you have weighed the costs and benefits so that you avoid making losses in the long run.

In the next chapter, you will learn how to set up a wallet and perform transactions through exchanges.

Chapter 6: Frequently Used Terminologies and their Definitions

If you are taking interest in Ethereum, it is essential to learn the terminologies and their definitions. Here are a few frequently used terms and their definitions.

Ð (Stands for ETH)

D with a stroke is a word used in Icelandic, Middle English, Faroese and Old English. It stands for "ETH". It is frequently used in words like Ðapp (decentralized application) or ÐEV. In these words, Ð represents Norse letter "eth". The uppercase Ð "ETH" symbolizes the Dogecoin (cryptocurrency).

Dapp (Decentralized Application)

It is a service operated without a main trusted party. This application enables direct communication, agreements and interaction between resources and end users without any middleman.

DAO

DAO is decentralized autonomous organization. It is a particular type of contract on blockchain that is enforced, automate or codify the working of a company, including expansion, spending, operations, fund-raising, and governance.

Identity

It is a set of cryptographically demonstrable interactions that have property created by a similar person.

Digital Identity

The particular set of cryptographically certifiable transactions signed by similar communal key describe the behavior of digital identity. In numerous scenarios of the real world, it is necessary that arithmetical identities overlap with identities of the real world. Confirming this without any violence is a mysterious problem.

Unique Identity

The particular set of cryptographically certifiable interactions has property that was created by a similar person with added constraints that a person can't have numerous unique identities.

Reputation

Property of a character that other objects believe that character can be either (a) competent at a particular task, or (b) trustworthy to a context, i.e. not to deceive others, even for a short-term profit.

Escrow

If two entities (mutually-untrusting) are involved in business, they may desire to pass funds through a 3rd party (mutually trusted party) and instruct this party to send funds to payees after showing the product delivery evidence. It is essential to decrease the risk of payee or payer committing fraud. Both the 3rd party and construction party is known as escrow.

Deposit

It is a digital property put into a particular contract involving a second party like if particular conditions are unsatisfied, this property will be automatically surrendered and credited to the counterparty as an insurance against conditions, or donated to a charitable fund or destroyed.

Web of Trust

If X highly rated Y and Y highly rated Z, then X is trusting Z. Powerful and complicated mechanism to determine the reliability of a particular individual in particular concepts may theoretically be collected from this standard.

Incentive Compatibility

Incentive-compatibility is a protocol if everyone is following the rules and doing well than trying to cheat, at least lots of people are agreed to cheat together at the similar time.

Collusion

In the scenario of an incentivized protocol, when numerous participants conspire (play) together to game the particular rules to their benefits.

Token System

It involves tradeable fungible virtual good. More officially, a token structure is a databank mapping addresses to figures with property that the key permitted operation is a transfer of N tokens from X – Y, with particular conditions that N is non-negative, N is lesser than current balance of X and a document approving the transfer is numerically signed by X. secondary "consumption" and "issuance" operations can exist and transaction fees are collected and instantaneous multi-transfers with numerous parties can be possible. Usual use cases are digital gift cards, company shares, cryptographic tokes in networks, and currencies.

Block

It is a data package that has zero or even more transactions, the hash of parent block (previous) and other data. The total blocks, with each block except for the early "genesis block" encompassing the hash of its previous blocks known as blockchain and have the whole transaction history of the network.

Keep it in mind that some cryptocurrencies (blockchain-based) use a word "ledger" as a substitute of blockchain and these two words are approximately equivalent, though in systems that utilize the ledger term, every block contains a copy of present state (for instance, registrations, partially completed contracts and currency balances) of each account allowing its users to dispose of obsolete historical data.

Dapp

Ðapp means decentralized application. It is pronounced as Ethapp because of the utilization of uppercase ETH letter Ð.

Address

An address of Ethereum represents one account. For EOA, it is derived as last 20 bytes of the communal key controlling an account, such as cd2a3d9f938e13cd947ec05abc7fe734df8dd826.

The hexadecimal format is based on 16 notation. It is indicated openly by affixing 0x to an address. Console function and web3.js accept addresses without or with this prefix, but their use is encouraged for transparency. Since 2 hex characters represent every byte of address and a preceded address contain 42 characters. Several APIs and apps are meant to implement new address scheme (checksum-enabled) in Ethereum Mist wallet as of 0.5.0.

Hexadecimal

It is a common representation layout for byte sequencing. It is beneficial because the values are epitomized in a particularly compact layout with the use of two characters in each byte (the characters are [0-9] [a-f]).

Ether

It is the name of Ethereum (a currency). This short name is used for computation within EVM. Obscurely, ether is a unit in this system.

EOA (Externally Owned Account)

It is an account controlled by one private key. If you have a private key linked with EOA, you can get the ability to send messages and ether from it. All contract accounts have a particular address. Contract accounts and EOAs may combine in a single type of account during Serenity.

Gas

It is the name of cryptofuel consumed while execitomh codes via EVM. The gas is a payment of performance fee for operation on Ethereum blockchain.

Gas Limit

The gas limit is useable for individual transactions and blocks (block-gas-limit). For each transaction, the gas boundary signifies the determined gas amount that you want to pay for the execution of the transaction. It is designed for the protection of users from getting exhausted while executing malicious or buggy contracts. The floor of gas limit increases with the introduction of the homestead from 3,141,592 gas – 4,712,388 gas (it is 50% increase).

Gas Price

The cost in the ether of a gas unit is specified in a transaction. After launching homestead, the default price of gas decrease from 50 Shannon – 20 Shannon (it is 60% decrease).

Web3

A Web3 paradigm is an important form that refers to a particular occurrence of improved connectedness between decentralization of applications, services, and all types of devices, semantic storage for online information and artificial intelligence apps for the web.

Epoch

It is an interval between every renewal of DAG utilized as seed by PoW algorithm Ethash. The epoch is defined as 30,000 blocks.

Cryptography (elliptic curve)

It refers to a particular approach to public-key cryptography on an algebraic arrangement of elliptic arcs over finite fields.

Wallet

In a generic sense, a wallet is anything that allows you to store ether or other tokens of crypto. In a crypto space, the wallet can be anything from an individual public/private key pair, such as single paper wallet to manage numerous key pairs like Ethereum Mist Wallet.

Contract

A piece of code on Ethereum blockchain to encompass executable functions and a data set. These functions implement with each Ethereum transaction for particular parameters. The input parameters decide the execution of functions and interaction with data within and other than a contract.

Mining

It is a process to verify contract execution and transaction on Ethereum blockchain in an altercation for an incentive in the ether with the help of mining of each block.

Mining Pool

Pooling of available resources by miners, who share processing power on a network and split reward equally, as per the amount of contribution to solving each block.

Mining Reward

The total amount of ether (cryptographic tokens) that is offered to the miner who mined a novel block.

Tate

It refers to a snapshot of data and balances at a specific point in time on the blockchain. It refers to the condition of a specific block.

Blockchain

It is an ever-growing series of blocks of data that grows with each new transaction. Every new block becomes the part of blockchain and chained to current blockchain through a cryptographic proof of work.

Business Terms

Here are a fee business terms that will help you to understand the difference between Ethereum and traditional terms.

Traditional Agreement (Deed Of Partnership)

It is a legal document important to write at the time of formation of a partnership to detail the rights and duties, profit share and important terms and conditions for each partner entering in the partnership venture. It is important to have a deed of partnership in order to avoid potential conflicts in future. It enables you to specify the division of profit between all partners according to their share. It is a legal agreement that helps you to avoid any disputes and differences in future. Deed of partnership should be signed by all partners and stamped by the concerned authorities of your state.

Flight Capital

Funds that are transferred to other countries in order to avoid high taxes or to fulfill the needs of a person are known as flight capital. It is a situation in which investors move their securities to other countries to avoid specific risks, unstable economy or political condition. Sometimes funds are transferred to other countries to get a higher return in the other country. Flight capital is common in those countries that have high inflation rate and offer a low return on the investment because of some certain economic conditions. Flight capital crops up when the assets and money quickly flow out of the country because of the unstable economy of the country.

Deficit Financing

When government expands more money than its revenues (collected through taxes) over an exacting period of time, this situation is known a deficit financing. Borrowing or minting of new funds is necessary during this process to cope with the difference of spending and revenue. Deficit financing shows the inefficiency of government, tax evasion activities and wasteful spending instead of devising countercyclical policy. There can be different reasons for deficit financing, but the influence of government deficit can be great on the national economy. Governments plan this expenditure in order to increase business activities to cover the shortfall by generating additional revenue.

Demerit Goods

Oversupplied goods in the society by the market are known as demerit goods. These goods are harmful to society because of their negative consumption externalities. Use of demerit goods may cause brim over effects on the third party without any compensation. An externality is a condition in which actions of consumers or producers can be the reason for negative or positive effects on the third party. Tobacco, alcohol, drugs, junk food, and prostitution are some examples of demerit goods. It is important to remove the leftover demerit goods from the market otherwise their excessive consumption can put negative effects on the health of consumers and society.

Depression

Depression is a stern and protracted recession in the economic activities. Depression is another name of severe downturn that usually lasts for two or more than two years. A considerable increase in unemployment, diminishing the value of the available currency, diminishing output, frequent rates of bankruptcies and sovereign debt defaults, reduction in trading activities and volatility in the value of currency are some characteristics of depression. Depression can lead to economy shutter down due to a decrease in investments and consumer confidence. Although, the recession is quite normal during a business cycle for few months depression can shut down all economic activities for a number of years.

Direct Marketing

Direct marketing is direct communication with a potential customer through the physical promotional material to convey important information about specific products and services. It does not involve any kind of online marketing or television advertisements. Different promotional materials like catalogs, mailers and flyers are used in direct marketing. It is considered as an effective technique because it removes the "middle man" for the promotion purpose and you can directly convey your message to potential customers. Direct marketing is a preferred way for those companies who do not have brand recognition and also could not afford other marketing techniques because of their smaller advertising budget.

Discount Loan

It is a type of loan that is issued at the desire of borrower and interests as well as other charges are calculated at the time of loan granting. Total of interest and other charges are calculated to deduct them from the face value of the discounted loan. In this type of loan, the borrower gets low amount as compared to the face value but he/she is still responsible to repay the full face value of the loan. The discount loan is a short-term loan and the borrower has to repay complete amount according to the given payment scheduled in installments. The borrower has to pay principal amount on an immediate basis.

Discretionary Order

It is an order that enables broker to delay the execution of a transaction at its prudence with an aim to get a better price for the customer. Discretionary orders are also known as not-held orders, and under these orders, the broker can take the decision about the sale and purchase of securities at best possible price to provide added benefits to customers. Some discretionary orders are restricted in terms to limit the amount of discretion for the broker. Discretion order is used to increase the price range, and the investor is liable to limit the discretion to the broker. A trader can decide the timing of buying and selling of securities.

Double-dip Recession

Double-dip recession is a recession followed by transitory upturn and another recession. It is a situation under which positive gross domestic product growth slithers back to the negative after a specific period of time. Reasons for a double-dip recession may vary in each economy but decelerate trend in the demand for goods and services because of layoffs and expenditure curtails from the previous downturn. Double-dip recession is a worst case for the economy because it can derail the economy and the economy can move back into deeper and long-lasting recession. It can worsen the situation and the recovery of the economy can be really difficult.

Economic Growth

It is an increase in the production capacity of the economy over a period of time. Economic growth is also referred as a rise in the market value of goods and services to yield more. Inflation is a real term that is used to measure the actual economic growth. GDP or GNP per capita is important to consider for the comparison of the economic growth of two countries. Technological changes have great role in the economic growth because the use of latest technology can increase the production rate in a short period of time. It is really easy now to spread the words about your precuts through the internet to grab maximum customers for your business.

Economies Of Scale

It is a cost benefit that rises with the rise in the yield of a product. Contrary association between quantity produced and per unit fixed cost can be the reason for the rise in economies of scale. Fixed cost shared over the production of a large number of goods can reduce the production cost and increase the number of goods produced. It is a better way to reduce per unit variable cost to provide direct benefits to the company. Economies of scale can affect different areas of a large enterprise including production and purchasing department. It can put positive impacts on finances of the organization.

Elasticity Of Demand

Calculation of association between the demand of a good and change in its price over a specific period of time is known as the elasticity of demand. It is an economic term that is used to define price sensitivity and its effects on the demand for a certain product. The elasticity of demand can be calculated as:

Elasticity of Demand = % Change in Quantity demanded / % Change in Price

Any small change in the price can affect the demand and responsiveness of particular products. A large change in price can decrease the demand for a specific product while any smaller change can cause a dramatic increase in the demand for the similar product.

Equilibrium Price

It is a balanced state in which market supply and demand comes at equal level leads to the stability of prices. Typically, excessive supply of certain products in the market can decrease the value of product and decrease in price will increase the demand for the same product. Balance demand and supply of the product brings state of equilibrium. Equilibrium prices indicate a situation in which supply and demands match with each other in balanced proportion. Period of consolidation or oblique impetus can bring supply and demand at equal level and this situation is known as the state of equilibrium.

Exchange Rate

It is a difference between the values of two currencies of two different countries.

It is a price of one currency in comparison to the currency of another country. In simple words, the exchange rate is a rate at which you can exchange your currency for the money of another country. The exchange rate can increase or decrease the value of the currency. For instance, the exchange rate of one euro is higher than one yen, and this situation can lower down the value of the yen in another country. Currencies such as Japanese Yen, British Pound and Euro are typically compared with U.S. dollars.

External Debt

Borrowing from foreign lenders for your country is known as external debt. Commercial banks, governments, and intercontinental monetary institutions are foreign lenders. Repayment of loans and interest will be made in the similar currency in which the loan was received. For instance, loan received in US Dollars will be repaid in the same currency.

In order to get required currency, borrower's government may increase exports and trading activities with the lender's country. Countries with weak economy may have to face a debt crisis because of their inability to repay external debt. It shows the inability of borrower's country to generate sufficient profit over a period of time.

Factory Price

In simple words, it is a price charged by a factory for the delivery of goods. It may be the amount of money that is required to purchase important items for the factory including gasoline, machinery, raw material etc. This term refers to the cost of goods at factory other than shipping costs and taxes. Factory price is a price that is quoted by the manufacturer for the pickup of goods from the gate of the factory. Shipping and other costs are not incorporated in the factory price. This price is charged by factory owners in order to generate some additional revenue.

Fast Moving Consumer Goods

These are rapidly sold goods at a comparatively low cost, therefore, these are also known as consumer packaged goods (CPG). Goods like soft drinks, grocery items, toiletries and other perishable items fall under the category of quickly moving consumer goods. The retailers can sell these goods in large amounts by keeping low-profit share.

Walmart, Metro Group, Johnson and Johnson, Colgate-Palmolive etc. are famous retailers of quickly moving consumer goods. These all are globally recognized companies with the largest share in the worldwide market. Fast moving consumer goods are typically low price items like cell phones, digital cameras, mp3 players, video games etc.

Financial Equity

Shares and other financial instruments of an enterprise are usually sold in the share market to increase capital and this process is known as the generation of financial equity. Financial equity raises the funds for business to perform different profitable operations.

Financial securities include IPOs "Initial Public Offerings", preferred stock, quasi-equity instruments, convertible preferred stock, shares, and warrants. These all instruments are treated as financial equity and can be sold to increase the capital of the business. Sale and purchase of financial equity are known as equity financing to fulfill financial needs of the company without taking a loan.

Fiscal Policy

It is an important government policy for the appropriate utilization of government revenue (taxation) and expenditure to bring considerable positive changes in the economy. This policy has great importance because it is used to control unemployment rate, inflation rate and to stabilize business cycle to bring stability to the economy.

Fiscal policy works on the idea of John Maynard Keynes, who thought that the governments have the ability to bring improvements in the bad condition of the economy by adjusting tax rates according to the expenditures of the government. The right application of fiscal policy can bring dramatic improvements in economic conditions.

Fixed Term Contract

It is a contract that is valid for an agreed period of time. An employment contract is an example of the fixed term contract. Under fixed term contract, an employee is responsible to perform his duties up to a particular period of time. Fixed term contract can be made for the completion of certain event or task. Employees under this contract can enjoy similar benefits over a fixed period of time just like permanent employees.

Fixed term contract worker is entitled to the same rights but for short period of time. The contract is designed to explain employment tenure, duties and responsibilities of employee and remuneration details.

Chapter 7: Smart Contracts and Ethereum Integration

Now that you know a bit more about how the Ethereum application operates, you should feel comfortable navigating the Ethereum dashboard and sending and receiving ether.This next chapter is going to get into how you can go about designing your own Smart Contracts on the Ethereum platform.To reiterate, a Smart Contract promises its users efficiency and censorship by only activating when certain parameters within a specific contract have been met.Let's take a look at how these Contracts work, and what they're all about.

Externally-Owned Accounts Versus Contract Accounts

The previous chapter focused solely on externally-owned accounts.In other words, externally-owned accounts are those that are able to send and receive currency.In an externally-owned account, a private encryption key protects the Ethereum wallet from being hacked, by essentially verifying that the sender did in fact intend to send the funds or messages that are being transferred.There is no overarching code that has been programmed into an externally-owned transaction, and the sender in an externally-owned transaction must sign off on the validity of the transaction through their private key before it is completed.This is done by verifying that the nonce number from the transaction matches the nonce number that the sender has sent the receiver.

Contrastingly, a contracts account is one that has a predetermined code already programmed into it.Once the code has been programmed into the contract, the contract itself can then receive messages based on the contract parameters.For example, let's say that you're trying to buy a home.There are many legal documents that must be acquired and signed prior to the mortgage being officially transferred to another person.Within Ethereum, you would be able to send messages to a mortgage contract.With each message that you send, Ethereum would use its storage space (through its swarms, remember?), and accumulate the messages for the contract.Once all the parameters within the contract have been met, the contract will then activate and be seen as completed.

While the previous example made it seem like a project can be completed within a single instance, it's important to understand that a Smart Contract is a working document.For example, let's say that you rent an apartment and your landlord has decided to use a Smart Contract for the lease agreement.You make your monthly payments to him/her with ether, and everything is running smoothly.One month, however, you are unable to make your rent payment on time because you had an expensive car repair for which you needed to pay.In fact, you don't have the money to pay your rent for a full seven days after the first of the month.Within the Smart Contract to which you agreed, there is a clause stating that if you do not make your rent payment within five days of the first of the month, you will be charged a $25 late fee.This means that once the sixth of the month comes, the Smart Contract automatically is able to recognize that you have not yet paid your rent and will automatically charge you the appropriate late fee.

While paying a late fee will more than likely be inconvenient for a tenant, the Smart Contract is able to make the job of the landlord much easier.A Smart Contract eliminates the human aspects of being a landlord that include having an awkward conversation with the tenant and demanding the late fee.Additionally, the landlord does not have the keep track of how many days have passed since the tenant was supposed to pay his or her rent.In this way, you should be able to see how a Smart Contract is a living document that does not simply stop working as soon as its parameters have been met.Instead, it acts more like an employee of a business or a clerk whose job it is to keep track of when people are in violations of their contracts.This can be an incredibly useful function.This example also demonstrates how a Smart Contract becomes the middleman that was a lawyer or a realtor in the past.The image below should be able to contextualize these examples a bit more concretely:

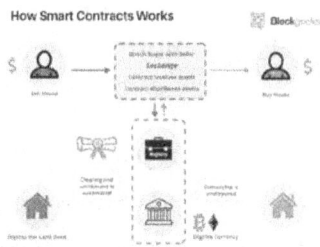

The Benefits of a Smart Contract

We've already discussed how Ethereum in general and Smart Contracts are able to reinforce trust with its users, and have also briefly touched on how Smart Contracts are able to eliminate potential middlemen.Those are two big benefits of Smart Contract implementation, but there are plenty of others that still need to be spelled out.These benefits include:

Duplicity: The public ledger on any blockchain system is an asset to a Smart Contract in the sense that you don't have to worry about losing your contract or having your information deleted.If something were to happen to your Contract, all of the other users on the network would see your Contract, and would be able to identify it as valid and true.

Speed: In the world before Smart Contracts, you would often have to wait for banks to process your funds or for paperwork to be processed by other legal entities.With a Smart Contract, your contract can be processed within minutes.

Avoiding Human Error: Human error often comes when many forms are being filled out by multiple people.With the code within a Smart Contract being written by one person, there is less human error altogether.

Safe: The blockchain system of cryptography that includes private keys, public keys, and ledgers, makes it difficult for a hacker to infiltrate a contract that's been put in place.This makes a Smart Contract arguably safer than a regular contract, because there are less eyes on what's going on and less potential for ill-willed human motivations to become involved.

Chapter 8: How to Mine Cryptocurrency

Before we talk about how to mine cryptocurrencies, we have to make sure you understand cryptocurrency transactions.

First off, we need to remind ourselves what we know already and add a little something extra to start you off.

We know that cryptocurrencies are digital currencies whose transactions are meticulously secured. These cryptocurrencies are stored in digital wallets that exist in a blockchain. We also know that these currencies are transferred from peer to peer directly. Third, we know that miners are responsible for validating these transactions and recording them in the blockchain.

When someone makes a transaction, the miners run the blockchains and use computers to calculate and tally the transactions. They do that to create updates of the transactions and ascertain the authenticity of the information, thus ensuring that all the transactions are secure and processed safely and correctly.

The miners then receive physically minted cryptocoins as payment for their services by the merchants or vendors of every transaction.

How Transactions Happen

Based on demand and supply, there is always fluctuation of the cryptocoins even though it as not fixed value. Sellers and buyers agree on a fair value based on the prevailing value of the cryptocoin in question before any transaction. The details of the transaction are then made public to all individuals in the blockchain network. However, the buyer and seller's identity remains concealed. Being a peer-to-peer transaction, there is no intermediary such as a banking institution taking part in this transaction and hence things that would otherwise be required such as credit card transaction fees are not part of the process.

Let's look at an example.

You want to buy an item whose value is $20,000. You discover that some seller accepts cryptocurrencies as a mode of payment. When you scout around, you find that the current exchange rate is $5,000 per currency (assuming you are dealing with bitcoin or a similar currency in this case). You get the seller's public bitcoin address from his/her website even though both of you are anonymous to each other. At this point, you can instruct the software on your computer or your bitcoin client to transfer four bitcoins from your wallet to the seller's address. Your bitcoin client will sign the transaction request electronically with your private key only known to you. Your public key, which is openly known, can be used to verify the information.

When your transaction broadcasts to the bitcoin network, the miners will then verify it in a couple of minutes. The four bitcoins will then be transferred to the address of the seller.

With that said, let's see exactly how you can mine cryptos.

How to Mine Cryptocurrencies

Mining cryptocurrency is simply the process of solving cryptographic puzzles with the use of specialized software after which you add various transactions to the blockchain (public ledger) in the hope of getting paid some coins as a reward.

From the above explanation, it is clear that with mining, you need software and hardware to confirm the transactions on the blockchain.

Is Mining Worth The Effort And Time?

As you probably know, mining cryptocurrencies is a kind of arms race that essentially rewards the early adopters. This means if you began mining Bitcoins back in 2009, you'd probably be thousands of dollars richer today.

In case you're wondering whether mining cryptocurrency is a worthy venture, you have to appreciate that if you're a low spender or you're just starting out, taking it as a hobby is better even though you can only earn a small income of a few dollars per day. Considering it a second income is usually not ideal or recommended because mining cryptocurrencies is not the most reliable way to make a significant amount of cash for most people. You'll only find the profit from mining cryptocurrencies becoming significant when you are willing to invest about $2000-$6000 in up-front hardware expenses.

If your aim is to generate good money right away, I would suggest you consider buying cryptocurrencies with money instead of mining them, and then tuck them away waiting and hope that they will increase in value like precious metals. If you just want to make a couple of digital bucks though, perhaps mining is the right path for you.

How do you do it?

The process of mining cryptocoins has to do with two functions:

Securing and verifying, which involves adding transactions to blockchains

Releasing new currency

As a miner, the individual blocks you add have to have a PoW or proof of work.

Mining also requires a good computer and a special program that will assist you, the miner, compete with your fellow miners in solving complex arithmetic problem. As you would expect, this requires large computer resources. You, on regular intervals, would try to solve a block that has the transactional data by use of cryptographic hash functions.

Therefore, the essence of mining is the idea of block rewards. The most coins go to the individual who gets a valid answer to the cryptographic hashing algorithm. This solution is an arithmetic calculation that uses the results of earlier block solutions so it is not possible to pre-calculate answers for future blocks without being aware of the solution to the preceding block. This history of transactions and block solutions is what essentially constitutes the blockchain.

In case you've forgotten, blocks are the files in which data associated with the cryptocurrency network is recorded. A block is therefore some kind of a 'page' of the 'ledger book'.

One block holds cryptographic signatures for the block and the transactions within it. You collect these transactions from the network, and they usually have a little fee attached (which becomes part of the block reward as well). A block also contains a difficulty value attached to the solution too, which can, over time, scale down or up; the goal is to maintain the rate of generating new blocks at a fairly constant level.

To make things easier for you, let's use bitcoin as an example.

Mining Bitcoins

To mine Bitcoins:

What you need

To mine Bitcoins, you have to have the mining software and hardware. In the past, you could effectively use the standard computer to mine Bitcoins before it shifted to power graphic cards. Currently, it is not possible to use such machines because of the growing difficulty.

Before you begin the mining process, you have to have a separate computer (usually known as the mining rig) to perform all the mining work. You can start by checking out the bitcoin miners on amazon to see the many types of miners spanning many different sizes and shapes. If you have the tech skills, you can also try building your own.

The idea is trying as much as possible to get the best rig here: acquiring a very competitive bitcoin miner. You should also consider visiting a bitcoin calculator such as the one offered here to help you assess the exact amount any particular miner will get you. These calculators are good because for one, they help you assess miners according to factors like profitability, which is usually the most important specification, and provide a crash course in other specs you may not be familiar with such as the difficulty factor, hash rate, and power cost.

For the hardware, you can also consider using Bitcoin ASIC chips that already have custom miners offering a performance of up to 1000x-10000 times the mining capacity from a computer or graphic cards. The antminer below is one of the best ASCIs on the market today.

Just remember that when you're starting out, you need to invest a substantial amount of money, be ready and willing to upgrade your equipment often and follow up on the bitcoin market conditions regularly (among other things).

Step 1: Download the bitcoin mining software

Once you get your miner, the next step is downloading the mining software. While there exist many out there, some of the most popular today include BFGminer and CGminer, which are inherently command line programs. You can go for EasyMiner if you prefer a GUI miner. If you don't know how to configure the command line programs mentioned above, please visit this guide.

Step 2: Join a mining pool

Once you have your hardware and software in place, you are just about ready to mine your bitcoins. However, you should join a bitcoin mining pool. The reason for this is simple:

Since mining rewards the miner who solves a puzzle first, and the chance that you will be the one to discover the solution is equivalent to the portion of the network's overall mining power, you will definitely stand minimal chance of discovering the next block on your own (because you obviously have a small share of the mining power).

Let's look at an example:

You are using one ASCI Antiminer; this means that the single ASCI antiminer would represent less than 0.001 percent of the mining power of the network. This is a very little chance of finding the next block, which means you would face a lot of difficulty going up. This also means the only solution is mining pools. When you work together in a pool and share the payouts amongst yourselves, it means each member will receive a steady flow of bitcoins.

Some of the top bitcoin pools around the globe you could join include the following:

AntPool

BTC.com

F2Pool

BTC.TOP

Step 3: Get a wallet for your bitcoins

A cryptocurrency wallet is the equivalent of a bank account. It lets you receive cryptos, store them, and send them to other people. We have two main types of wallets:

Hot wallet: A hot wallet is a type of crypto wallet you install on your own mobile device or computer; with it, you have total control over your coins' security. However, given that they are on a device connected to the internet, they are obviously quite insecure.

Then we have the 'hardware wallet'. This one maintains increased security levels to protect coins by storing them offline, which keeps them (along with data), out of hackers' reach.

Some of the best hot wallets you can get include the following:

Bread Wallet

Exodus

Mycelium

Armory

Hardware wallets include:

Trezor

Ledger Wallet Nano

Step 4: Start mining!

By now, you have everything in place. Connect your miner to a power outlet and boot it up, making sure you connect it to your computer. Open the mining software and probably enter your mining pool's password and username (if you joined any). Having configured that, you will begin mining for bitcoins.

Please read this guide to make sure you meet the other prerequisites, especially if you're doing solo mining.

Now start the .sh file if you're using Linux or .bat file if you're on windows.

Congratulations! You are now mining cryptocurrencies!

You can check the performance of your miner through the software itself or through the pools website.

Use the formula below to calculate your profits:

P=M-W

Where;

P is profit

M is the mined value- this means if you mine 2.3 BTG that is probably valued at $150 USD, the value for M should be 345.

W is the power consumption or wattage.

The result should thus look something like:

P= 345-67

P=278.

Your monthly profit therefore is $278.

Chapter 9: The Ethereum Enterprise Alliance

Fortune 500 companies, blockchain startups, and research groupsgot together in the spring of 2017 and started da nonprofit organization that is called the ethereum enterprise alliance.Some of the companies that are in the enterprise are Microsoft and Intel. There are at least a hundred and sixteen members that are part of the enterprise alliance.

The official EEA logo

This alliance was created as an open source for private blockchains that these companies are using. The blockchains will address the data that is stored in the blocks for every sector that is using blockchain, which means that everything from banking to entertainment is being used by these companies. The alliance has found a solution that addresses the ethereum ecosystem that is used by ethereum users, including the alliance. The technology that the alliance is creating is not only helping them achieve their goals, but it is making ethereum more efficient for everyone who uses it.

Some alliance members have announced that they are creating new projects on their blockchain. One of these projects is a hybrid architecture which will bridge private chains to public chains. The new blockchain will be public and will open up new possibilities that were not available before if it had not been for the company putting their project ideas into action.

The blockchain for ethereum is always changing which means that the information in the company's blockchain will be open for the public to see what the company is doing. So, at the end of the day, Ethereum Enterprise Alliance wants to open up private blockchains to free blockchains.When you open up the closed blockchains, what the businesses are doing is not going to be hidden anymore, and the public will see how they are operating from the inside. There may not be a way for the public to interact with the company, but that does not mean that the public is not going to be able to better understand enterprises. And, if the public can see everything, then how are the big fortune 500 companies going to be able to hide from their public when questions are asked about how their money is being used in creating a better experience for their customers.

One of the most significant sectors that wants to bridge the gap are the financial companies because they do not want to lose their customers. This is why they are wanting to join in with the blockchain technologies so that they can show their customers that they are willing to join in with the way that the future is going and continue to hold the trust of their customers.

It is always easier to understand when you can see that a company's ideas are actually put into action. So, let's look at a few companies that are working to put their ideas into action for the ethereum users.

JP Morgan Chase started a blockchain that stood between public and private blockchains so that payments could be sent and received. The inspiration came from a regulator that needed access to the transactions for their company while keeping their customer's privacy secure.

The Royal bank in Scotland announced that they were created a tool that would help clear out settlements on the distributed ledger. The ledger worked with smart contract technology and let users write out their settlement deals.

Chapter 10: Uses of Ethereum

Ethereum is a platform that allows the use of smart contracts and distributed applications on its blockchain. This makes it highly versatile and usable for many purposes. Of course, with its token known as ether, it can also function as any other cryptocurrency that works as a substitute for money. But, Ethereum has taken this a step further. With the use of smart contracts and distributed applications, the Ethereum platform can do lots of things, such as create other platforms, games, messengers, function as the base support of other altcoins, and many others.

As we have already discussed, Ethereum is composed of a vast computer network. There are thousands of computers connected to it. This is what makes it a decentralized network. This means that the computers that form the Ethereum network cannot be controlled, shut down, or manipulated by any entity, group, or individual. This, together with the use of smart contracts and distributed applications, opens the door to various uses and applications of Ethereum. Here are some examples:

Transactions

Ethereum can be used for the exchange of things of value. You can be sure that it will be risk-free. With the use of smart contracts, a transaction will only be executed and completed once certain conditions are met. This can also lower your cost involved in a transaction as it effectively removes any third-party service in a transaction. Ethereum has an IFTTT logic in its system. This refers to the "IF This Then That" logic. Once certain conditions are met, then execution of the contract can be made. Since it effectively removes the middleman in a transaction, then you can save on cost, especially in the long run.

Security against hackers

Ethereum can be used to protect you from hackers and other online attacks. As you already know by now, the Ethereum blockchain has a high security. It is virtually impossible to break into it. Hence, you can rest assured that your any sensitive data can remain protected.

Health records

Another use of the Ethereum blockchain is the recording and storage of hospital records. This can be used, especially when developing vaccines or in the process of treatment. For example, you can easily have a check-up in Canada and be treated in the U.S. easily since you have all your records stored in the Ethereum blockchain. This can also be used by doctors and hospitals in sharing valuable information with one another.

High-technology innovations

Ethereum can be used in the creation and development of high-technology innovations, such as self-driving cars and even self-piloted aircrafts. The possibilities that can be harnessed by the Ethereum platform are definitely way beyond how an ordinary cryptocurrency like bitcoin works. After all, Ethereum is far more than a cryptocurrency that functions as a medium of exchange. Instead, it encourages and drives the development of technology, which makes it open to limitless possibilities.

Effective storage of data

Regardless what kind of data or information that you want to store, you can rest for sure that it will be protected and secured when you use Ethereum. You no longer have to depend on a few servers and worry that they might get hacked or broken. With Ethereum, you can store data in thousands of computer networks spread across the world. With Ethereum, you can encrypt and send data to millions of servers very quickly.

Casino gambling

There are many cryptocurrency gambling sites today. You no longer have to drive just to visit a casino. You can gamble online in the comfort of your home using ether. There are also many live casinos that will allow you to play casino games with a real and professional dealer.

Profitable investment

It is not a secret that majority of those who own ethers use them as a form of investment. In fact, many people these days are so eager to learn about cryptocurrencies with a hope that they can make money out of the cryptocurrency market. The good news is that this is possible. In fact, there are already many people around the globe who have earned millions of dollars, and some have even attained financial freedom by investing in cryptocurrencies like Ethereum. Investing in Ethereum can be a highly lucrative investment. Just in 2017, the price of Ethereum in the market has increased by around 13,000%, and that translates to 13,000% profit had you made an investment in the same year. Take note that this does not include its many other price increases in the previous years. Here is the good news: If you take a closer look at the past and future trend of Ethereum, it is not hard to recognize that it still remains to be a highly lucrative investment today. In fact, many experts agree that the price of Ethereum will most probably increase significantly this 2018 to the point that it might be able to overtake Bitcoin and soon be the number one cryptocurrency in the market.

Additional Notes

There are, of course, many other uses and applications of Ethereum. Its platform is like a rich soil on which you can grow many wonderful trees. This will depend on how you use it, especially with respect to the use of smart contracts and distributed applications on the Ethereum blockchain. It should also be noted that although Ethereum has already gained a worldwide popularity and positive reputation, it is nonetheless still a young technology. This means that it still has a big room for improvement and growth. Now, this is actually a good thing, especially if the developers behind this cryptocurrency truly excel in what they are doing.

Chapter 11: How to buy, sell and store ethereum

There are a few ways to buy Ethereum, Ether or ETH (it'strading ticker name), and this guide is going to show you how you can get hold of some in the easiest way possible. Just like Bitcoin there can bequite a few cumbersome hoops to jump through but hopefully we can show you the best method to suite yourself. You can buy it with fiat currency, buy it with bitcoin or you can mine it.

Buy Ether with Fiat Currency:

There are a number of fiat exchanges that you can use to buy ethereum, or ether, directly with local fiat currency such as dollars euros or yuan. Kraken is a particularly good way to buy ethereum with dollars, pounds or euros and we've written a guide on that here. There are other exchange that you can use to get hold of ethereum immediately.

Buy Ethereum with Bitcoin

If you can't find the right currency to trade into or you think the bid ask spread is too high or there isn't enough liquidity you will have to buy it another way. The best way to do this is to buy bitcoin first.

One of the least hassle ways to buy bitcoin is to use Coinbase. Unfortunately Coinbase is only supported in afew countries and their pay by debit card feature is limited to a certainquantity per day.

Once you have found the right place to buy some bitcoin.... buy some. There are some loopholes to go through usually that can make the process painful, but that is only Companies trying to comply with legislation on KYC and AML.

Mine Ethereum:

Another option to get hold of some ether is to to mine Ether yourself. Alternatively, you can also try buying a cloud mining contract with Hashflare or Genesis Mining.

HOW TO BUY ETHEREUM / ETHER CLASSIC

Most people try to mine Ethereum but then get frustrated with the high up-front costs. People want Ethereum, so the easiest way to get Ether ends up being by simply just buying Ethereum tokens or Ethereum Classic.

Note: Before you buy Ethereum make sure you have a secure place to store your Ether! An Ethereum hardware wallet is the most secure option although not free.

Buy Ethereum with Credit Card or Debit Card

There are many ways to buy Ether with a CC and this section will discuss the 3 best options. Note that the fees will be around 3.5% for most options.

Coinbase

In the USA, Europe, Canada, UK, and Singapore, Coinbase is the easiest way to buy Ethereum with a credit card.

The fees will amount to 3.75% and you can buy instantly.

Note that Coinbase only sells Ether and not Ethereum Classic. To buy Ethereum Classic you need to buy bitcoins on Coinbase and transfer the bitcoins to Poloniex to exchange them for Ethereum Classic.

CoinMama

CoinMama is a nice option because you can buy less than $125 worth of Ethereum without the need to verify your identity. You can instantly sign up and buy Ethereum. You can have Ether in your wallet within about 20 minutes.

The downside to CoinMama is the fees are pretty high and come out to around 7%. CoinMama supports credit and debit cards from basically any country.

Note that CoinMama only sells Ether and not Ethereum Classic. To buy Ethereum Classic you need to buy bitcoins or Ethereum on CoinMama and transfer the bitcoins to Poloniex to exchange them for Ethereum Classic.

CEX.io

CEX is a Bitcoin broker based in the UK and now sells Ether as well as bitcoins.

You will need to complete an extensive ID verification process. Once done you can buy Ether for about 4% fees.

Buy Ethereum with Bank Account

In the USA and most of Europe, Coinbase is probably the easiest way to buy Ethereum with your bank account. Some other options are BitPanda, Kraken and Gemini.

Buy Ethereum or Ethereum Classic with Bitcoin

You can buy Ethereum or Ethereum Classic with Bitcoin at nearly any crpytocurrency exchange. This is because most of the global Ethereum trading volume is actually done in the ETH/BTC pair, and not the ETH/USD pair.

You can buy either Ethereum or Ethereum Classic with Bitcoin or any other crypto using Changelly.

Poloniex has the most trading volume globally so you can get the best exchange rate there although there are 0.15-0.25% fees.

Using GDAX you can easily buy Ethereum for no fees, although there is less liquidity than at Poloniex.

Buy Ethereum with PayPal

Unfortunately, there is no easy way to buy Ether with PayPal.

You will first have to buy bitcoins with PayPal on VirWoX. Once you've purchased bitcoins there you can use an exchange like Kraken or Poloniex to exchange your bitcoins for Ether.

Chapter 12: How Markets are changed by Cryptocurrencies

Over the last couple of months and years, several countries around the world have legalized cryptocurrencies and effectively rendered them legitimate payment methods. Others such as Poland, Germany, Switzerland and Australia have passed legislation that supports use of cryptocurrencies.

This means in future that trading in the markets could accept payments in cryptocurrencies such as ethers and bitcoins. It could also mean that stock exchanges, traders, brokers and investment banks could adopt blockchain technology. This will make trades open and verifiable while easing payments and reducing costs of conducting business.

Cryptocurrency firms are now raising money using means such as ICOs. An ICO is simply an initial coin offer is a method of raising funds by offering cryptocurrencies to investors at discounted prices. This is a new way of raising funds even though it is very similar to IPOs. Investors will have a new tool to add to their portfolios.

Many institutional investors have now started adding cryptocurrencies to their basket of investments. Cryptocurrencies are in some cases viewed as commodities which can contribute to an investor's portfolio and increase profitability. Common cryptocurrencies like Ethereum, bitcoin and even Litecoin are already being actively traded online.

The cryptocurrency market is currently worth over $145 billion. This represents a huge market that cannot be ignored or wished away. It is changing the understanding and outlook of the economic and financial sectors. Just this year (2017) the value of 1 Bitcoin surpassed that of gold. The growth of this market has grown exponentially and is far much bigger than most people would have imagined.

Blockchain

The blockchain is bound to affect the way the markets operate. Consider the stability and reliability of this platform. It supports most cryptocurrencies, ensuring that all transactions are open, verified, anonymous and indelible. All these are qualities that desirable in the finance world.

There are currently many companies working to develop blockchain based applications for use in the finance sector. Stock markets, for instance, can use the blockchain technology to ensure that trades are verifiable and accountable and to prevent rogue traders from mischief. This could introduce new levels of integrity and openness that customers desire.

The likeliness is extremely high that blockchain technology will soon replace current monetary systems. At the very least, they will play a parallel role similar to traditional stores and online stores or print media and digital media.

Current systems used by banks and other financial systems are complex and quite risky, exposing customers and institutions to unnecessary risks and high operational costs. Blockchains such as the Ethereum-based one use smart contracts. Critics view them as inflexible but many observers love the fact that they are very secure and resilient to attacks even as they are not influenced or affected by government policies.

Cost savings

Cryptocurrencies have drastically reduced the cost of transactions. One way they have achieved this is through elimination of intermediaries and middlemen. Most intermediaries charge fees and slow down processes. When they are eliminated, process become cheaper and faster.

The blockchain also eliminates the need for back offices where hundreds of people are employed. If the back office can be eliminated, institutions will save money and transactions will be cheaper and hence more affordable. The speed of processing transactions will also increase significantly.

It is an undeniable fact that cryptocurrencies and the blockchain are changing the landscape in the economic and finance sectors. They have introduced new ways of transacting and better, more reliable platforms which tend to reduce costs, save on time and make processes more efficient. It is expected that as the months and years go by, more and more solutions will be adopted that will make use of both the blockchain and cryptocurrencies.

Chapter 13: Roadblocks

All new things have a development stage. Knowing what's hindered people in the past will help expedite your learning process and get you into the market quicker. Let's go over some of these roadblocks.

Scalability is an issue that is within the mainstream payment networks because there are about 2000 transactions a second. You will need to change the block size limit so that there are more supportable transactions per second. It is possible at the current stage that if Ethereum were to increase in size dramatically, there is an issue of finite nodes being able to be used.

Some transaction blocks require over half the network hash power to reverse. This is partially due to the high-security needs and could be remedied if reversals cost the initiator a fee.

Stamping is an issue currently as well. Generally, in the blockchain, blocks are made every single day. However, creating blocks more frequently causes the payment systems to become incredibly slow. Thus the network works most efficiently with a specific number of blocks made per day. Additionally, since each user is given a specific time stamp (distributed along a normal curve), and thus no nodes are ever within 20 seconds of each other, thus making communication between nodes purposefully delayed.

There are other roadblocks, but the developers are all working on streamlining every step of the process within Ethereum and smart contracts. Patience is a virtue when waiting for developers to fix problems.

Chapter 14: Tips on Investing in Ethereum

If you find that mining is not something you want to try your hand at there are still other options for making money with Ethereum. As a matter of fact, the majority of people usually prefer to invest or trade their Ether to make a profit rather than to do all the work and put all their money into mining.

When compared to mining, investing in Ethereum is the more popular choice for a number of reasons. But the question you should always ask yourself is whether or not investing is the right choice for you. The answer will depend largely on your level of risk tolerance. Because cryptocurrencies are very new, they tend to be considered as high-risk investments, but as history has shown, they can sometimes be an investment that can pay off well in the future.

If you want to compare numbers, many are buying into Ethereum simply because Bitcoin's price has soared to such high levels that it is out of reach of many people. One thing for sure, investing is a waiting game, and you will need to have a lot of patience as you wait it out in hopes of making a profit.

If you plan on investing in Ethereum there are a few things you should keep in mind:

1. Ethereum has intrinsic value; because it has a purpose in the virtual world, it is of value to that world. This means it is viewed differently than Bitcoin, which is used more as a trading tool than to transfer commodities of value.

2. To realize your profits earned from your investment in Ethereum, you will need to at some point to sell or trade them and transfer the value back into your local currency. With that in mind, make sure that you earn enough in the sale to pay your capital gains taxes you will accumulate from your sale.

OK, now that you've made the decision let's get into some simple ways to invest in Ethereum and make money while you do.

To do this there are three fundamental functions you must be able to do: buy, sell, and trade.

Tricks to Buying Ether

One of the first things you will need to do is to buy your Ether. Remember, Ether is the fuel that makes Ethereum work. Technically, you can't buy Ethereum (it is the network that people trade on), but you can buy Ether to give you access to the network. Just like when you go to an amusement park they usually give you tickets or tokens to ride the rides. Ether is the token that lets you on the ride called Ethereum. Without it, you cannot play the game.

It's important to remember that it is not considered a currency in the same way as Bitcoin and can only be used on the Ethereum network. You won't be able to buy or sell any products or services with it so to make a profit you will need to develop certain strategies that will allow you to manipulate situations to your advantage based on the price movements.

Up until now, we've discussed Ether, but there is another cryptocoin also associated with Ethereum, the Ethereum Classic. You need to know the difference between the two as they have different rates and different values. Ether, the original blockchain technology for Ethereum carries more weight and more value while Ethereum Classic, an offshoot of the original has yet to gain its footing. For the sake of profitability potential, we'll discuss the tricks to buying Ether.

How to Find the Right Exchange House

#7. Of all the decisions you make on investing in Ethereum, choosing the right exchange house is crucial. If you're in a hurry to get your Ether, then you're likely to want to take a few shortcuts. It could be tempting to simply do a Google search and pick the first exchange that appears on your list, but you might be doing yourself a disservice if you do. There are several important factors that you need to weigh carefully to make sure that you get the exchange that will give you the best chance at earning a profit on your investment.

To buy Ether, you will have to convert your fiat currency (dollars) into Ether. One of the best places to do this is through a money exchange. When you consider that cryptocurrencies are not backed by governments, you can understand why they are easily used globally. So, just like when you travel to another country you have to convert your currency into the kind that is used in the country you're visiting, you'll have to do the same to get Ether.

Now, in a new country, you might find an exchange house in every community, to find an exchange house for Ether you need to go online. One of the most popular exchanges for buying Ether is coinbase.com. However, they are not the only one. Other exchange houses like Kraken, CoinMama, and BitPanda are also popular choices. The exchange house you choose will depend on where you are, your method of payment, and the requirements they may set for investing. Here are a few things you need to ask before making a final decision.

Is the money safe: This is important especially if you're planning on investing large sums of money. While cryptocurrency is more resistant to hacking than other more traditional systems, it doesn't mean that there is not someone out there trying to be the first to break through its barriers. To that end, you want to find out what steps the exchange is taking to prevent these types of cyber-attacks from happening. What methods have they put in place to protect your assets from being stolen?

By its very nature, once a transaction is made in the crypto world, it is permanent and irreversible. Therefore, any exchange that doesn't have security precautions in place makes your investment more vulnerable to lose than anything else.

Some exchanges have implemented something called Fund Security, which works much like the FDIC does when dealing with the banks, but others have not. Before you choose to use an exchange find out how they plan to protect your investment.

Transaction Fees

It is not enough to have a good price when it comes to buying Ether. Exchanges houses have a number of fees they also charge for each transaction made. This is how they make their money. Just like your stockbroker charges a commission when they make a trade, the exchange houses do too.

Exchange house fees can be classified in two ways: Maker Fees or Taker Fees. Maker Fees are those imposed when you boost the liquidity of the books and Taker Fees occur when you lower the liquidity.

In addition to those fees, there may be other surprise fees that could eat away at your profits. Fees for withdrawals, deposits, or small deposits all can hit your profits pretty hard. When looking for the best exchange house, look under all of the rocks to find those hidden fees and determine how they are going to impact you.

If you're planning on buying and holding for a long time you may not have much to worry about as you won't accumulate a lot of transaction fees but if you're looking to enter a buy low sell high game of get in and get out, you definitely want to know what your actions are going to cost you down the road.

Location

It may be easy to conclude that location is not important since all exchanges are already housed online but you'd be wrong. Their location could actually impact how they protect their funds. Exchange houses in the US for example, are governed by US standards and regulations. This means that they are required by law to protect your assets in some form or another. However, an exchange located in China where there are no regulations can fold at any time taking your money with them, and they would not be required to make any attempt at helping you to recover your funds.

Another issue with location has to do with the predominant form of currency found in that area. If you're in the US and dealing with Euros, you will incur an additional fee for converting your money back to US dollars, which will definitely cut into your profits over time.

Language also should be a factor. If an exchange is based in Japan, their support staff may only be available in Japanese so if you have a problem that needs to be resolved it could become that much more complicated.

Because exchange houses are the primary tool for buying and selling Ether, it is important that you take the time to choose the right one. The more you invest in researching these houses, the better chance of you finding one that will meet all of your needs so that you can make the most out of your investment.

#8 How to Pay for Ether

Once you have found the exchange house, your next step is to pay for your Ether. Just like with everything else, there are several ways to pay, so it makes sense that you find a method that is more suitable to the way you plan to do business.

Buying Ether with Cash: Caution is warranted when you choose to buy Ether with cash. Not only is there a high level of risk when it comes to validating that you're making a legitimate purchase, but it also requires you to have a face-to-face meeting with strangers, a step that often puts people at risk. It is strongly recommended that you only purchase Ether this way from people you already know and trust or make sure that you meet in a public place where the level of risk can be reduced.

Buying Ether with Your Bank Account: It is simple to buy Ether with a bank account transfer. It is one of the best ways to buy large amounts without hassle. However, it should be noted that when you buy large quantities of Ether this way the transaction process could be considerably slower than other methods of purchasing.

Buying Ether with Your Credit Card: One of the easiest ways to purchase Ether is with a credit or debit card. This can be done through any one of the exchange houses found online. Unlike many smaller digital currencies, Ether is readily available in almost every exchange house you want to work with.

Tricks for Trading Ether

By now, you have probably gathered that when it comes to dealing with Ethereum, safety precautions are always a key. While we may not be dealing with a life or death situation, making a mistake when dealing with any type of cryptocurrency yields a permanent result. Unlike with traditional currency where you can call the bank to resolve a mistake the very nature of cryptocurrency does not allow room for that. So, when it comes to trading your Ether, there are several things you must keep in mind to protect your assets.

#9. Pay attention to the details:

While trading can be very rewarding, there is also a level of high risk. You need to always be on your toes and give 100% of your attention to the deal that is being made. By doing so, you can protect yourself from all sorts of negative situations that may arise.

#10. Know Why You're Trading:

You need to have a strategy that starts with you knowing your ultimate goal for the trade. Not all trades are made for profit so be on the lookout. Just like in the stock market, there are often people who buy up huge quantities of stocks and methodically sell them off to unsuspecting buyers at ridiculous prices. Buyers who are not paying attention to the market trends or the present situation will inevitably lose out. These vultures are literally lying in wait, looking for those who are not very savvy about the market and are simply too anxious to get their foot in the door. Always know what your goal is and do your research before making a trade. There are times when it might be better to do nothing than to rush into a decision without knowing all the facts. Remember, with each trade your Ether is exposed to potential losses so never rush to a decision.

#11. Target and Stop:

For every trade, you should first have a specific target in mind. This is a figure you pre-set for taking a profit. If you buy Ether at $350/coin, you might want to set your target or goal for profit at $400/coin before you are willing to trade it for another altcoin. On the other hand, you should also have a stopping point set for cutting your losses in case the prices start to fall. So, if you buy your Ether at $350/coin and the price starts to fall you could set your stop at $300. This way you don't run the risk of losing all your money on the investment. In essence, you are stopping your losses.

To do this, it is important to try to take the emotion out of the deal. If you're a trader that falls in love with the idea of trading or wants to hang on to the very last minute in hopes of a turnaround, you might not be in the right game. Setting up your Targets and Stops beforehand will help you to focus on the end game better rather than pushing the envelope when things don't go so well.

#12. Control Your FOMO:

FOMO (fear of missing out) can be scary. When you see the price going up, up, up and you haven't traded yet, your emotions can quickly take control over your power of reason. When you have a plan, you are far less likely to be emotionally involved in your trade.

#13. Manage Your Risks:

To trade profitably, it is best to look at the small gains that will eventually accumulate into big ones. The bigger gains usually don't last and can fall as quickly as they rise.

Another way you can manage your risk is to diversify. Never put all of your eggs in one basket. Instead, when your portfolio has a variety of coins, you increase your potential for gains. Never have more than a small percentage of your portfolio in high-risk gains, investing the bulk of your funds into stable coins like Bitcoin and Ether is your best chance of gaining you a steady profit.

#14. Trade in Relation to Bitcoin:

While you may not be trading Bitcoin, every altcoin including Ether has some sort of relationship with Bitcoin. When Bitcoin's value rises, altcoins begin to lose their value, and when Bitcoin's value loses momentum, then the altcoin currency tends to become more volatile.

At times it will be difficult to predict what the future is ahead so when that happens, keep your targets close and your stops even closer.

#15. Expect a Loss in Value:

If you plan to trade for the long term, it is only realistic to expect that there will be periods where the Ether will lose some of its value. But even here, knowledge is power. Even though Ether is one of the leading coins in cryptocurrency, don't neglect your homework. It is important that you follow the chart's and take the time to find the high and low as well as the periods of stability. Once you know these periods, you'll be able to anticipate times when the trend will start increasing again and you can make your profit.

#16. Learn How to Manage Fees:

It is always wise to avoid buying from the order book or the Taker. By buying this way, you can rack up quite a few more fees than normal. For example, on the Poloniex exchange, the difference between the Taker fees and the Maker fees can be as much as 0.1%, which is quite high. Since all exchanges make their money through fees, you need to learn your way around them and strategize ways to cut the costs so you can keep more of your profits.

#17. Look for Optimal Conditions to Trade:

Never start a trade until you are sure you are in the best position for yourself. You should have a plan on how to get in and when to get out. Without a plan, the pressure of the trade could get to you, which almost always leads to losing money. If you're ever in doubt, it is best to wait for a better opportunity than to take an unnecessary risk.

Two Common Trading Strategies

#18. Buy and Hold:

When it comes to trading Ether, two trading strategies are most often used. The Buy and Hold Strategy simply means that you'll buy the Ether and hold onto it until the price reaches a point where you've made a profit.

#19. Active Trading:

The Active Trading strategy is a little more difficult to achieve results. It requires that you set pending orders and put stop losses in place. Part of the challenge in this strategy is that you must find the right exchange. Not all exchanges will allow you to set these in place so when you're doing your research on exchanges if you plan to do active trading, you will need to find out which exchanges will permit this kind of trade.

If you are working with an exchange that does not permit this, it doesn't mean you cannot do any active trading. To accomplish your goals, however, it will be necessary for you to set up alerts that will keep you apprised of market movements. With alerts, you will receive messages that tell you that your mark has been reached or is about to be reached and then you will have to place your trades manually.

To set up alerts, several online websites monitor the movements of cryptocurrencies and will send you notifications based on your stop loss or buy goals. These do charge a fee, but if you're planning on making money as an active trader, this could pay for itself in a very short period of time.

#20. Learn how to analyze the movements of Ether:

It can be exciting to feel like you're getting in on something that has as much potential as Ethereum but it is also important to keep in mind that Ether, like all other cryptocurrencies, it is still in its infancy. That means that while there is huge potential for making money, there are also many risks.

In the stock market, you are encouraged to do a technical and fundamental analysis to determine the trends and movements of a stock, but because Ethereum is so young, there is not much in terms of reference points for such a study. With that in mind, do as much research as you can but focus on managing your risk. It is a highly volatile market and without a measurable history to relate to, it could explode into something even bigger tomorrow, or it could plummet in price.

That said, probably the best decision you could make when buying Ether is to buy low and sell high. For now, it is probably the best technical advice anyone can give.

If you are interested in a more fundamental analysis of Ether, keep in mind that the continued success of Ether will obviously rely more on the quality of the apps that will be used on the platform. That means you will need to keep abreast of what new things are being introduced on Ethereum every day.

The best way to follow these activities is to tap into the accounts, blogs, and chat rooms that are constantly abuzz with new Ethereum ideas. By keeping a close eye on the Ethereum Twitter account, Ethereum

Blog, and any other pages dedicated to monitoring Ethereum, you'll know when something big is about to happen, and you can get in on the ground floor.

There is no way to list all of the best strategies, tips, and tricks that people use to make money with Ethereum but at least now you have the basics. You are now ready to enter the market and start making money. Just remember, the more you know, the safer you are, but there are no guarantees that you will turn a profit, but the potential is extremely high.

Chapter 15: How to Mine Ethereum

You can use any PC to mine Ethereum, but never use light devices with underpowered GPUs (such as laptops). Ethereum mining using the CPU is not practical. It takes a longer period to complete the process, but the revenue is lower compared to expenses.

When it comes to mining, GPUs are better than CPUs – and GPUs are at least 200 times faster than a typical CPU. Do note though, that nVidia cards are slower compared to AMD cards. If you're familiar with these cards' mainstream applications, you're probably wondering why AMD's offerings are considered superior. After all, nVidia is deemed better in almost every application there is. This is due to the fact that the main mining program for Ether is implemented in OpenCL – a software framework fully supported by AMD's GPUs. While nVidia GPUs can still work with the open-source framework, they're actually optimized for CUDA (the company's own version of OpenCL).

Indeed it takes a high volume of electricity in order to mine Ether. If mining is executed efficiently, an increase in profits can be gained by selling Ether tokens.

To be sure of your plan's profitability, you might need to use specialized mining calculators. There are also Ethereum Mining mini-computers for determining profits.

The Ethereum Mining Process

Below is a simple process to set-up your Ethereum mining node and begin mining your first Ether token.

Step 1 - Download Geth

The first thing that you need to do is establish your communication channel. You need to download a software called Geth. This will create a secure connection to the Ethereum network across the world while it coordinates your hardware. It'll provide updates on the processes being done, particularly those that need response from your end.

Geth is often downloaded as a Zip file, which you'll have to extract somewhere. It is ideal to use your C drive for this step. Use the Windows search option to look for CMD. If you are not certain, then you can browse around the search listing.

Step 2 - Locate Geth

The placeholder for the username is often similar to the system name (e.g. C:\Users\Username>). In locating Geth, you need to type in cd/ in cmd. This is a command to shift directory.C:\> should be on highlight, which shows that you are in drive C.

Step 3 - Create an Account

At this point, you should start creating your user account. To do this, start by making a call: type in geth account new then press Enter. C:\>geth account new should be displayed now in the command prompt. This step also involves setting up your password. You need to be cautious in setting your password, and be sure that you use a strong combination of alphanumeric keys. After keying in the password, press Enter and you now have a new account.

Step 4 - Download Ethereum Blockchain

Linking the Geth to the Ethereum network is needed before it becomes functional. Type in geth — RPC on the command terminal before pressing the Enter key. This activity begins with downloading the Ethereum blockchain and linking with the world's blockchain. This action is time consuming and relies on the size of the blockchain. It also relies on your web connection speed. Be sure to wait until it's done before proceeding.

Step 5 - Install Ethminer

Now, you need to install a program that will allow the GPU to operate the blockchain algorithm needed in the Ethereum network. The ideal option for this difficult task is Ethminer. Create a fresh terminal for command then access the terminal icon (active) that is located on the taskbar before accessing the terminal window in the menu.

In the newly opened terminal window, put cd prog then press tab. C:\>cd prog must appear in the window, press tab to show C:/> cd "Program Files" then press Enter to display C:\Program Files>.

Type cd cpp and press Enter in order to proceed to the Ethereum mining folder. The terminal window will show C:\Program Files\cpp-ethereum> right after you press tab again.

Step 6 - Start Mining

Key in Ethminer –G on your terminal window then press the Enter key in order to start mining with your GPU. This will start Ethereum mining after the Directed Acyclic Graph (DAG) gets created. It's a huge file mainly kept in your GPU's RAM so it can be compatible with ASIC. Nonetheless, you have to ensure that your HDD has sufficient space before doing this step.

If necessary, mining Ether using a CPU is also possible. Just type ETHMINER then push the Enter key to initiate the mining process. The creation of DAG is necessary in this phase, after which the connection with Ethminer will be taken over by Geth.

Chapter 16: Uses Of Ethereum

Ethereum may be a virtual currency in that it doesn't have a physical form but like other currencies, it can be used for real-life applications – and change the way that business is Conducted

Applications

Here are some of the applications that could benefit greatly from Ethereum.

Health Care Services

Ethereum will change the health care system in a major way. Using the system, all patient records can be stored, accessed, and shared by all hospitals. This can lead to major breakthroughs in the field of medicine and vaccines. As a patient, you can have your checkup while you're on vacation in Singapore and the doctor there will have all your medical records. If you have one of those smart watches that record your heart rate and blood pressure that can also be linked to the hospitals providing even more information about your current health status.

Online Transactions

The economy runs on transactions. Ethereum will transform transactions significantly. Inside the system, smart contracts can be created. These contracts make it possible for anything of value to be exchanged completely risk-free. Actions based on agreements are executed automatically, avoiding the need for an intermediary so there will be fewer fees to pay. For example, you want to buy a photo from Shutterstock. You can transact directly with the photographer.

Gambling

In the US alone, the gambling industry's worth is estimated to be at $240 billion. As you may already know, casinos (physical or online) sometimes employ shady and fraudulent activities to get more profit and they can get away with this because of the lack of transparency. These activities will be exposed if Ethereum is used since all records are publicly visible.

Benefits Of Using Ethereum For Transactions

Security/Preservation of personal information

Ethereum works in a decentralized mode, so no central server (e.g., a website) can be attacked by a hacker to get your personal information.

Privacy/Anonymity

You use search engines to look for the information you need. You may not know it but these search engines collect personal information from you and then sell the data to advertisers. That's why you get search lists that are related to your previous searches and preferences. Although this can be convenient, it's also a form of privacy invasion. Using Ethereum's blockchain technology, a log will be recorded every time a search engine utilizes your data and makes this log available to the public. Advertisers will then have to be careful with how this data is used or risk exposing their illegal activities.

In Politics

We practice democracy when we vote during election day. But election results can be easily altered. Ethereum technology makes results transparent and visible to the public ensuring more transparent elections.

Reliable Data Storage

Cloud storage services such as Dropbox store subscribers' data in server farms. A server farm is a network of computers and most of these machines are located in a single location, making it vulnerable to data loss due to natural occurrences or terrorist attacks.

Using a decentralized storage system, data will be stored in thousands or even millions of computers making it more resilient. Ethereum's blockchain technology can encrypt and transfer between networked machines at a fast rate.

Chapter 17: Ways to Make Money with Ethereum

Now that you have laid the groundwork for your Ethereum investment, you are ready to get started. For most people, the idea of investment is to buy low and sell high. This is a basic and fundamental rule that will carry you a long way. However, this is not the only way you can make money with Ethereum. In fact, you'll be surprised to know just how many different ways money can find its way to you through Ethereum. However, for this chapter, we're going to give you the three primary methods that have worked well for investors so far. Mining, Investing, and Trading.

Mining

Probably, one of the most important factors in the Ethereum Blockchain is the mining process. While you do need to know your way around a computer to be a miner, statistics show that earning money as a miner is actually more profitable than direct investing when you do it right.

The challenge, however, is that to mine you may need to invest in some pretty special equipment that may set you back a pretty penny. While you can mine with a CPU or a personal computer processing unit, the speed at which it moves could cut the amount of money you can make.

GPUs, on the other hand, tend to work as much as 200 times faster than your personal CPU and the faster it works, the better your chances of making money.

You will also need to download specific software to get started. So whatever system you plan to use, you want to make sure that it will be compatible with Ethereum's network. The good news is that you no longer need to download the full Ethereum Blockchain (now more than 20GB) to get started, nor do you need to deal with those cumbersome line miners with manual instructions anymore.

That said, you will need to invest something in getting your mining system setup. If you're not willing to shell out for quality hardware, your odds of generating a good income will be seriously curtailed. Still, the potential for profits is very good now. Below is a list of some pretty good reasons why people have chosen to mine Ethereum.

It can serve as a great way to subsidize your current income

It's an easy way to obtain Bitcoins. You can trade Ethereum for Bitcoin.

Ethereum is a great entry coin opening the door to the cryptocurrency market.

You can have your own voice on the Ethereum network through the mining process.

These are not the only reasons why you might want to consider mining but just a few to start you thinking. Because of the payout for every block solved, the idea of mining for Ether can be very tempting. However, it is a very technical approach to entering the cryptocurrency market and is not for everyone. If you are interested in learning more about the mining process and how it works, we'll go into more detail in the following chapter.

Investing

By and large, the majority of people who want to enter Ethereum's world will be investors, and this is for a good reason. Of all the money making ventures concerning cryptocurrency, investment is probably the most direct approach.

Those who choose to be investors realize that it is a game of numbers, but it is also psychological game as well. Learning how to predict what the public will do in any given situation can be an interesting endeavor, especially when it concerns money.

Successful investors are those who are diligent when it comes to studying charts, keeping up with the news reports, following social media, and keeping abreast of all sorts of things that might have an impact on the price of a coin. They are not the people who check the exchanges every day to see how the price has changed but they are those who are content to look beneath the surface to find out why the price has changed and to determine what might happen in the future.

Statistics show that Ethereum's price has increased 5,700% in the past year. The novice investor looks at those figures and says wow! I want to be a part of that and jumps into the fray without checking further. The savvy and cautious investor wants to know what caused the price increase and whether or not Ethereum will be able to maintain it and grow from there.

If you're trying to decide if investing is the best option for you, there are four different investment personalities; see if you fit into any one of these categories. Understanding your personal investment personality can help you to decide which investment strategy is going to work best for you when you're ready to put your money down.

The Preserver

The Accumulator

The Follower

The Independent

The Preserver: This type of investor is not a huge risk taker. His concern is in maintaining what he already has. The preserver closely monitors his assets and is very conscientious about protecting their losses. He is slow to make decisions mainly because he wants to avoid the possibility of making mistakes that could cause him to lose money in the end.

If you feel that you are a preserver, then you will be more interested in long-term investment strategies. When looking at Ethereum over an extended period of time, you will notice that while the price fluctuates a great deal in the short-term, the general trend is upward giving you a better chance of profit in the long-term.

Short-term investments for the preserver are risky because of the volatility of the market. Prices may see extreme lows and highs all in the same day. Preservers tend to be very emotional when faced with any type of loss and may be inclined to pull out of an investment too soon and lose out in the end.

The Accumulator: The accumulator is more interested in amassing wealth. They are more likely to be in control of their investment decisions and are willing to take risks. These are the people who study the graphs and charts diligently and are very confident in whatever conclusion they come to. Most accumulators have already seen success in other ventures of their lives and can easily transfer their experience in analyzing situations and coming to the right conclusion to their investment practices.

One problem the accumulator often faces is his overconfidence. Because they have had a history of success, they approach investing in cryptocurrency with the idea that they will always make the right choice. The result is that with the cryptocurrency market, there is extreme volatility, so the odds of miscalculating are even higher than with any other type of investment tool.

The Follower: This is the person who doesn't have a lot of confidence in his ability to gauge the market, so he tends to follow the advice of others. He is also easily swayed by exciting news or interesting stories about what a currency is doing. The follower rarely has any ideas of his own and generally depends on others to tell him which strategy he should take.

His greatest challenge has to do with the following the crowd mentality. He will readily pool his money with others simply on the basis of the idea that someone else says its good. This can present a major problem because often, by the time he hears about a profitable investment opportunity, it is already too late to take advantage of it. He may get in on an upswing in prices just before it begins to fall again. This strategy often leads to losses that could affect your overall plan.

The Independent: The independent investor is just like it sounds. This is the person who has his own ideas and strategies and is not afraid to steer away from the pack. These people are much more interested in the process of investing and can study the market in minute detail. They are very analytical and critical, and they have the confidence to stand on their own in any decision they make.

Their greatest challenge is their confidence. They tend to rely too much on their own thinking process and are not as willing to listen to and accept the advice of others, which can cause them many problems and often a loss of money in the end.

With each investor personality, there are always advantages and disadvantages. As long as you understand what type of investor you are and what your negative drawbacks are, you can create a plan of action that will help you to overcome the obstacles that are naturally a part of your particular investment personality.

Trading

Most people are guilty of using the words "investing" and "trading" interchangeably. And while there are many similarities in the thought process, they are very different tools.

If you've done any trading on the stock exchange, you will likely be very familiar with the art of trading Ethereum. You will need to become very adept at analyzing charts, and predicting future price movements. The trader must always do extensive research so that he can make well-informed decisions, and avoid giving in to the temptation to react to the volatility of the market.

However, that is where the two begin to diverge. When trading on the cryptocurrency exchange, your goal is to concentrate your efforts on buy and sell orders. When you place an order to trade, you are predicting that the price will go up or down and your payout comes when the market price reaches your prediction.

Successful traders have very specific characteristics in their personality. First, they are able to look at a chart and a graph and see beyond what's written on the page, they are able to see the big picture, connecting the past with the future to identify potential trading opportunities.

Second, they are very logical thinkers and are skillful analysts. They look at every detail with a critical eye and are capable of seeing possibilities where many others cannot.

They are extremely organized, are able to make decisions quickly, and are capable of following through despite what others may think or say. It takes time to become a good trader. You have to commit to a daily ritual of trading and studying charts and graphs consistently to develop that critical eye that will see all the possibilities that may not always be hidden in full view. The most successful traders are those that have developed their strategic skills and are good at planning. They are not only able to predict market movements with a high rate of accuracy, but they are emotionally balanced enough to make quick decisions under all sorts of market conditions. Whether the market is moving up or down, they know exactly how to make the most of the movement so they can achieve financial success.

Whether you decide to become a miner, investor, or trader, the possibilities of making money with Ethereum are very good as long as you recognize the risks and do your research. Those who choose to rush into the market and try to take shortcuts will definitely suffer the consequences.

Of course, these are not the only ways one can make money with Ethereum, but these are the most common and the most successful. So, now that you understand your personal investment style, you know exactly how you want to make money with Ethereum, let's take a look at these options in a little more depth.

Chapter 18: What Is Ethereum Proof of Stake

If you have not yet heard about PoS, we are going to update you on what is happening in the world of Ethereum. Apparently, Ethereum wants to change the consensus that is in its distribution network to what is now called, proof of stake. We are going to explain what this means, and how it might affect you.

Proof of stake

For us to understand the proof of stake, let us look at the proof of work (PoW) which is the current system that Ethereum uses. Now, during the transfer of Ethereum, miners solve a puzzle in a blockchain. This blockchain utilizes lots of computational power. Once you form a blockchain, a reward that is in the form of a transaction reward is awarded to you. But it also depends on how fast you can come up with a solution to the puzzle.

When it comes to proof of stake (POS), this whole process will go away because puzzles won't be necessary anymore. The element that needs puzzle solving is removed, and the reward awarding process is altered. Instead of showing people your speed of hash rate calculation, you are required to prove the number of Ethereum that is in your possession. This is done with the help of a master code. When the master code is created, a lockup has to be done on certain amounts of Ethereum to prove ownership, and depending on the proof of stale you own; the rewards will be distributed. You can create several master nodes that have lots of Ethereum in it, and you can earn more in this process.

Since we already have learned the definition of proof of work and proof of stake, we are going to look at the main difference about each concept. It is obvious that PoS is cheaper, efficient and faster. But what is its cost?

The major difference is in the way the methods handle untrusted communication in the network. When it comes to PoW, is a user decides to cheat during block creation, other nodes forgive the dishonest but when it comes to PoS, the user is penalized for being dishonest. Since there is forgiveness in dishonesty when in PoW, there is nothing that restricts dishonesty. But in PoS, everyone makes sure that they are not culpable of dishonesty, to avoid any penalty.

Where do you come in?

Proof of stake as we have seen in other currencies like in dash is an instance where mining does almost 50% of the rewards, and the rest is the proof of stake. Proof of stake is advantageous in many ways. A major benefit is that you don't have to use computational power to solve math problems. Another advantage is the lockup feature. When Ethereum is locked up, scarcity happens which makes the price to rise.

The year 2017 is the year where this project has to be completed, and we hope that the Ethereum development team have to ensure that the code is stable and also provide support to the miners. In case the miners are not supported, then Ethereum will be broken up as it has happened before, something we hope not to see again.

The good thing is that Ethereum has set a strict time for all this to take place. In case the switch is not done, it will all be a complete disaster. We shall see how the progress turns out in the next few months.

As much as the proof of work is going to be eliminated, miners should not worry about what happens to them, because there are many more other cryptos that can be mined. For instance, one can mine a profitable crypto called Zcash when using AMD GPUs. This is an exciting phase for everyone.

POS is not that much of a perfect solution since those who own a lot of ETH have a huge advantage over those who are just learning about it, but it is a great step in the right direction. It is part of the growth curve, and we can only wait and see what happens in the next few years in Ethereum.

Chapter 19: PRIMARY PURPOSE OF ETHEREUM

Unlike Bitcoin, the primary purpose of Ethereum is not to act as a form of currency but to enable "smart contracts" between the parties without forcing them to trust or use a middleman. Smart contracts are computer codes that can facilitate the exchange of money, property, content, or anything of value. Because these contracts run on the blockchain, they run just as they are planned without any possibility of downtime, censorship or fraud.

Ethereum enables developers to build decentralized applications (or Dap.) Because these computer programs are made up of code that runs on a blockchain network, they aren't controlled by a central entity.

Think of Ethereum as the world's first decentralized virtual supercomputer.

How Does Ethereum Work?

Here is an example that can help you understand how Ethereum works. When you send a message via Whatsapp, the message goes through the Whatsapp data centers. In other words, the content of your message gets in the hands of one big company. This system is called a centralized network of computers.

When you send a message using a decentralized application, on the other hand, the messages is sent to a network of independent computers all over the world owned by regular people. Every computer does a bit of the work and receives rewards in the form of a digital asset called ETHER.

HOW TO MAKE MONEY WITH ETHEREUM

• The technology that underlies Ethereum means that it can be used for a number of other purposes that will be built off of a decentralized and autonomous system. Simply put, it potentially will be a revolutionary technology with the potential to impact a whole spectrum of industries.

• As the demand for the Ethereum platform and its smart contracts enabled network increases, the value of Ethereum as a cryptocurrency will continue to surge.

• Stability – Ethereum had an organic growth, without massive spikes, and it seems to be stable, if not even predictable. The increasing demand and value of a certain cryptocurrency serve as an indicator of its potential. Whatever the reason is, it still increases the demand – meaning a further increase in Ethereum price.

• The developers at Ethereum want one to think of the network as a large virtual computer that facilitates applications to run. It is indeed this allure that has been the reason that the project has got the backing from a number of individuals such as Bill Gates.

• Making waves in the established industry: Microsoft offers Ethereum as a blockchain-as-a-service.

• The second biggest market cap after Bitcoin only, and one of the most popular cryptocurrencies ever in terms of volume.

Chapter 20: CRYPTOCURRENCY SECURITY

The security of cryptocurrencies is two part. The first part comes from the difficulty in finding hash set intersections, a task done by miners. The second and more likely of the two cases is a "51%" attack". In this scenario, a miner who has the mining power of more than 51% of the network, can take control of the global blockchain ledger and generate an alternative block-chain. Even at this point the attacker is limited to what he can do. The attacker could reverse his own transactions or block other transactions.

Cryptocurrencies are also less susceptible to seizure by law enforcement or having transaction holds placed on them from acquirers such as Paypal. All cryptocurrencies are pseudo-anonymous, and some coins have added features to create true anonymity.

CRYPTOCURRENCY LEGALITY & TAXES

Bitcoin Taxation

While cryptocurrencies are legal in most countries, Iceland and Vietnam being an exception – Iceland mainly due to their freeze on foreign exchange, they are not free from regulations and restrictions. China has banned financial institutions from handling bitcoins and Russia, while saying cryptocurrency is legal, has made it illegal to purchase goods with any currency other than Russian rubles.

In the U.S., the IRS has ruled that Bitcoin is to be treated as property for tax purposes, making Bitcoin subject to capital gains tax. The Financial Crimes Enforcement Network (FinCEN) has issued guidelines for cryptocurrencies. The issued guidelines contain an important caveat for Bitcoin miners: it warns that anyone creating bitcoins and exchanging them for fiat currency are not necessarily beyond the reach of the law. It states:

"A person that creates units of convertible virtual currency and sells those units to another person for real currency or its equivalent is engaged in transmission to another location and is a money transmitter."

Miners seem to fall into this category, which could theoretically make them liable for MTB classification. This is a bone of contention for bitcoin miners, who have asked for clarification. This issue has not been publicly addressed in a court of law to date.

Cryptocurrency Services

There are a host of services offering information and monitoring of cryptocurrencies. CoinMarketcap is an excellent way check on the market cap, price, available supply and volume of crypto currencies. Reddit is a great way to stay in touch with the community and follow trends and CryptoCoinCharts is full of information ranging from a list of crytocoins, exchanges, information on arbitrage opportunities and more. Our very own site offers a list of crypto currencies and their change in value in the last 24hrs, week or month.

Liteshack allows visitors to view the network hash rate of many different coins across six different hashing algorithms. They even provided a graph of the networks hash rate so you can detect trends or signs that the general public is either gaining or losing interest in a particular coin.

A hand website for miner is CoinWarz. This site can help miners determine which coin is most profitable to mine given their hash rate, power consumption, and the going rate of the coins when sold for bitcoins. You can even view each coins current and past difficulty.

THE FUTURE OF DIGITAL CURRENCIES

Probably the most important point to note about cryptocurrencies is the distributed and decentralised nature of their networks. With the growth of the Internet, we are perhaps just seeing the 'tip of the iceberg' in respect of future innovations which may exploit undiscovered potential for allowing decentralisation but at a hitherto unseen or unimaginable scale. Thus, whereas in the past, when there was a need for a large network it was only achievable using a hierarchical structure; with the consequence of the necessity of surrendering the 'power' of that network to a small number of individuals with a controlling interest. It might be said that Bitcoin represents the decentralisation of money and the move to a simple system approach. Bitcoin represents as significant an advancement as peer-to-peer file sharing and internet telephony (Skype for example).

There is very little explicitly produced legal regulation for digital or virtual currencies, however there are a wide range of existing laws which may apply depending on the country's legal financial framework for: Taxation, Banking and Money Transmitting Regulation, Securities Regulation, Criminal and/or civil law, Consumer Rights/Protection, Pensions Regulation, Commodities and stocks regulado, and others. So the two key issues facing bitcoin are whether it can be considered as legal tender, and if as an asset then it is classed as property. It is common practice for nation-states to explicitly define currency as legal tender of another nation-state (e.g. US$), preventing them from recognising other 'currencies' formally as currency. A notable exception to this is Germany which allows for the concept of a 'unit of account' that can therefore be used as a form of 'private money' and can be used in 'multilateral clearing circles. In the other circumstance of being considered as property the obvious discrepancy here is that, unlike property, digital currencies have the capacity of divisibility into much smaller amounts. Developed, open economies are generally permissive to digital currencies. The USA has issued the most guidance and is highly represented on the map below. Capital controlled economies are effectively by definition contentious or hostile. As for many African and a few other countries the topic has not yet been addressed.

Starting from the principles of democratic participation it is immediately apparent that bitcoin does not satisfy the positive social impact component of such an objective in so far as its value is not one it can exert influence over but is subject to market-forces. However any 'new' crypto-currency may offer democratic participation when the virtual currency has different rules of governance and issuance based upon more socially based democratic principles.

Conclusion

Thank you again for downloading this book!

I hope this book was able to help you to gain a better understanding of Ethereum and cryptocurrencies in general. Hopefully you now have an understanding of how you should go about trading and investing this promising new cryptocurrency.

The next step is to get started with Ethereum as it has incredible potential!

Thank you and good luck

CPSIA information can be obtained
at www.ICGtesting.com
Printed in the USA
BVHW041706210621
610126BV00010B/2106